IMAGES OF WALES

SOUTH WALES COLLIERIES

VOLUME TWO

The Ballad of Trehafod

Words and Music by Hawys Glyn James

Old Tre-haf-od, sum-mer dwell-ing, Crys - tal clear its brooks and falls,

Lush, green fields and pleas-ant wood-land, Pure air filled with wild bird calls, Fish-es

fill - ing pond and riv - er, Sheep and cows on hill-sides green, Scatt-ered

here and there a farm - house, Charming, slee-py past-oral scene. Black

go - old, Black go - old, Blue scarred min-ers dig-ging for black gold.

2.
Down the village came the drovers,
Cattle lowing on their way,
Down the lane past Hafod Ganol,
Travelling by, no time to stay.
Through Trehafod on to Cardiff,
Came the monks with flocks of sheep,
Passing by Yr Hafod Uchaf,
On they trudged o'er hillsides steep.

3.
Coal-shafts purchased in mid
                    seventies,
Coed Cae then the Hafod shaft,
Lewis, W.T. the buyer,
Coal produced by men's hard graft,
Next the Bertie, then the Trefor,
Sunk by W.T. for gain,
Named in honour of his two sons,
In the Park these shafts remain.

4.
Stack, big wheel and whirring winder,
Gone the trees and fields of green,
Dust-filled air and chugging jiggers,
Black-faced miners on the scene,
Ponds and streamlets dark and grimy,
Clank of coal-drams in the air,
Pick and lamp replaced the
                    ploughshare,
Birds' song hushed by hooter's blare.

5.
Zion, Bethel and Bethesda,
Penuel, Wesleyan too,
And Saint Barnabas, Siloam,
Great religious fervour grew.
Pubs and clubs in competition,
Boxing, brass band, fish and chips,
Football, horse shows, fairs, cymanfa,
Sunday School and seaside trips.

6.
Nineteen-fifty-six explosion,
Down the Lewis Merthyr mine,
Sad the hearts in old Trehafod,
And the number killed was nine.
Then the village links were severed,
When in nineteen-eighty-three,
Lewis Merthyr's pit-life ended
When they closed the colliery.

7.
Rhondda's story is recorded,
Through the folk who had a dream,
And a clear conscious vision,
To present the Black Gold theme,
And make certain that the hardships,
Toil and sweat have left their mark,
So that lives of Rhondda people,
Are remembered in the Park.

IMAGES OF WALES

SOUTH WALES COLLIERIES
VOLUME TWO

DAVID OWEN

The History Press

Acknowledgements

Thank you all for the wonderful stories, songs, poems, drawings and photographs of the South Wales Collieries, which have been given to me by the people from the mining villages of South Wales.

These have come from the early days of the coal industry through to the new millennium. I dedicate my book to the people of South Wales the Land of Song and in memory of all the miners who worked at the collieries.

I sincerely thank everyone for their kindness and help.
David Owen
Author and Archivist

Cydnabyddiaeth

Diolch am yr holl storïau, caneuon, cerddi, darluniau a ffotograffau aruthrol o Faes Glo De Cymru, sydd wedi eu cynnig i mi gan bobl pentrefi glofaol De Cymru.

Mae'r cyfraniadau yma yn dod o ddyddiau cynnar y diwydiant glo trwyddo i'r milflwydd newydd. Rwy'n cyflwyno'r llyfr yma i bobl Gwlad y Gân De Cymru er cof am y glowyr a wethiodd yn y pyllau glo.

Rwy'n diolch yn ddidwyll i bawb am eu caredigrwydd a cymorth.
David Owen
Awdur ac Archifydd

Front cover photographs:
Top: Lewis Merthyr Collieries in 1903. *Left to right*: Coed Cae, Hafod, Trefor and Bertie Pits.
Bottom: Lewis Merthyr Colliery First Aid Team, National Winners at Blackpool June 1954. *Left to right, back row*: Mr E. Pugh, Mr J. Jean, Mr John, Mr W.H. Newbury. *Middle row*: Mr H. Davies, Mr T. Wilson, Mr M. Rowley, Medical Secretary, Mr D. Davies, Mrs A. Harries, Mr H. Fisher, Mr G. Jones, -?-. *Front row*: Team Captain, Mr Joseph Reynolds, NCB Area Medical Officer, Dr D. Milne Dunbar, Colliery Manager, Mr Dick Richards (a former Powell Duffryn Employee), Dr Colin Evans, Area First Aid Officer, Mr W. McTiffin, Mr W.J. Jones, -?-.

First published in 2002 by Tempus Publishing
Reprinted 2005

Reprinted in 2010 by
The History Press
The Mill, Brimscombe Port,
Stroud, Gloucestershire, GL5 2QG
www.thehistorypress.co.uk

British Library Cataloguing in Publication Data.
A catalogue record for this book is available from the British Library.

ISBN 978 0 7524 2393 7

Typesetting and origination by Tempus Publishing.
Printed and bound in England by
Marston Book Services Limited, Oxford

Contents

Preface

Within the history of coal mining world wide, the enormous contribution of the Rhondda Valleys has never been equalled in its dramatic rise to industrial dominance in the production of steam coal.

Many major coal mining giants spring to mind, none greater than the Lewis Merthyr Colliery, Trehafod. Its history, from the 1850s to eventual closure in 1983, is in itself a formidable subject and I am proud to be associated with this book which not only records its huge influence on Rhondda's industrial expansion, but, that of course, an enterprise of this magnitude, with a workforce of thousands throughout its history would inevitably contribute in a substantial way to the rich cultural excellence associated with Rhondda.

My personal recollections begin in 1953, starting work underground at the Bertie Pit and becoming a member of the Lewis Merthyr Colliery Brass Band. In 1989 I joined with others in establishing a Centre wherein Rhondda's proud heritage would take its rightful place in the annals of mining history in the South Wales Coalfield.

Ivor England
Tour Guide

Rhagair

Yng nghyd-destun Hanes Glofaol Rhyngwladol dyw cyfraniad anferth Cymoedd y Rhondda heb ei guro yn nhermau twf syfrdanol i safle blaengar wrth gynhyrchu Glo Rhydd.

Mae yna nifer o gewri yn dod i gof, dim un yn fwy na Glofa Lewis Merthyr, Trehafod. Mae ei hanes o'r 1850au i pan gafodd ei gau yn 1983, yn destun arswydus a rwy'n arbennig o falch o fod yn gysylltiedig â'r llyfr. Mae'n adlewyrchu cymaint oedd y pwll wedi dylanwadu ar ddatblygiad diwydiannol Rhondda ond hefyd sut roedd menter o'r fath, gyda gweithlu o filoedd, yn gallu cyfrannu i'r profiad diwylliannol cyfoethog sydd mor gysylltiedig â'r Rhondda.

Mae fy atgofion personol yn cychwyn yn 1953, yn dechrau gweithio ym Mhwll Bertie ac yn dod yn aelod o Fand Pres Glofa Lewis Merthyr ac yn 1989 yn ymuno gyda eraill i sefydlu Canolfan lle gellir arddangos a chofnodi diwylliant balch y Rhondda a'i le priod yn Hanes Mwyngloddio Maes Glo De Cymru.

Ivor England
Dywysydd

Foreword

Rhondda Valley, probably the most famous coalmining valley in the world, for the mining of coal and corresponding wealth, has now gone, but its fame lives on. The valleys were once an environmental black spot due to the mining activity that dominated the area since the turn of the last century. Now transformed, the area boasts lush green, picturesque landscapes, which are a delight to see and enjoy.

I congratulate author David Owen on his first-rate and impressive collection of photographs in this book presenting a pictorial history of the Lewis Merthyr Collieries, Trehafod Village and the Rhondda Heritage Park in the South Wales Coalfield.

The past is our inheritance and the photographs in this book are reminders to keep our bygone times alive and to preserve the rich and proud heritage of the valleys of South Wales. Future generations will be able to browse through these pages and see the changes in the colliery and the villages that surround them.

The Rhondda Heritage Park is a living testament to the mining communities of the world-famous Rhondda Valleys and offers insight into the rich culture and character of the Valleys in a unique, entertaining and educational environment for all ages.

David Morgan
Friends of the Rhondda Heritage Park Chairperson and former Mayor of the Rhondda

Rhagair

Cwm Rhondda, yn sicr y dyffryn glofaol enwocaf, lle mae mwyngloddio glo a'r cyfoeth a ddaeth ohono wedi diflannu, ond mae enwogrwydd y Rhondda dal yn parhau. O safbwynt yr amgylchedd roedd y gweithgaredd mwyngloddio, oedd wedi dominyddu y ganrif ddiwethaf, yn niweidiol. Erbyn hyn mae'r tirlun wedi newid, gyda phorfeydd gwyrdd a golygfeydd hardd, sydd yn bleser i'w mwynhau ac i'w gweld.

Rwy'n llongyfarch David Owen ar ei gasgliad anferth o luniau yn y llyfr yma, sydd yn cyfleu hanes Glofeydd Lewis Merthyr, Pentref Trehafod a Pharc Treftadaeth y Rhondda ym Maes Glo De Cymru.

Ein gorffennol yw ein treftadaeth ac mae'r lluniau yn y llyfr yma yn cadw ein hatgofion cyfoethog a balch o gymoedd De Cymru yn fyw. Fe fydd cenedlaethau i ddod yn gallu pori trwy'r tudalennau i weld y newidiadau yn y Glofeydd a'r pentrefi yn yr ardal.

Mae Parc Treftadaeth y Rhondda yn dystiolaeth weledol o fywyd cymunedau byd enwog Cwm Rhondda ac yn cynnig i bob oedran adlewyrchiad o ddiwylliant a chymeriad Cwm Rhondda, mewn amgylchfyd unigryw, adloniadol ac addysgiadol.

David Morgan
Cyfaill Parc Treftadaeth y Rhondda, Cadeirydd a chyn Faer y Rhondda

Introduction

As we enter the twenty-first century, the world of Lewis Merthyr Colliery in its heyday at the turn of the twentieth century seems to be literally just that – another world. The change of the South Wales Valleys from rural calm to intense industrial activity, epitomised by the story of the Rhondda Valleys, is surely one of the great stories of human endeavour – producing the massive wealth for the mine owners and the tremendous hardship that was endured by their workforce in creating that wealth.

And yet it would be inaccurate to suggest that all was doom and suffering because, at times within this 150-year story, mineworkers were considered to be in a fortunate position, enjoying well-paid employment and housing with better amenities than was the norm for their time. The hardships also produced tremendous community spirit, the beginnings of the 'welfare state' and were a major driving force in the creation of the trade union movement.

The story of Lewis Merthyr Colliery, the Rhondda and its people also underlines a great paradox – if the life was so harsh, why did the communities fight so hard to retain it? The answer may lie in the fact that poverty and the unknown were an even worse option; the 'knowns' of community, 'butties' (friends) and camaraderie were somehow comfortable, no matter how dangerous and tragic a life it sometimes became.

Today the story is told at Rhondda Heritage Park and this book traces the story 'From Pit to Park' in pictures and words. The Park was formed because it is important that this story is told; it is part of our heritage. It would have been sacrilege had the world's most famous coal mining valley passed in to history without a living history museum and tourist attraction to tell its story. We have visitors here from all over the world as well as from down the road. The Park attracts many tourists into the area who spend thousands of pounds in the local economy. It has great value as an educational resource and many schools and colleges visit.

The Park owes its existence to the Friends of Rhondda Heritage Park, Rhondda Cynon Taff County Borough Council and its predecessors, the European Union, the Welsh Development Agency, Wales Tourist Board and others who helped in its formative years.

And what of the future? The Rhondda and the other South Wales Valleys have changed significantly in the last twenty years, a change almost as dramatic as their industrialisation. Community spirit remains strong, challenges of social deprivation and the mobility of the workforce still face us, but 'hi-tec', 'multi-media' and service-based industries now present new opportunities. The Valleys are returning to their green past, heron fly along the River Rhondda, trout swim in its waters – unthinkable twenty years ago when the river still ran black. We can surely look forward to a better future and Rhondda Heritage Park must be part of that future, it too must move with the times and look forward as well as back.

For the benefit of future generations, it must continue to tell the story.

John Harrison
Director of the Rhondda Heritage Park

One

Lewis Merthyr Collieries in the South Wales Coalfield

The name Rhondda is often quoted as being synonymous with that of coal. However, just over a century and a half ago the Rhondda Valleys were almost unknown and existed as a sparsely populated rural wilderness.

For centuries the Rhondda Valleys remained in pastoral glory, with clear running streams and waterfalls, and beautiful trees and flora. The small sheep-rearing community that populated the few scattered farmhouses, existed as they had for centuries, as a sleepy rural community.

A remarkable change took place in Rhondda in the second half of the nineteenth century with the discovery of coal. By the end of the century Rhondda was one of the most important coal producing areas in the world. At its peak the coal industry in Wales employed one in every ten persons and many more relied on the industry for their livelihood. Rhondda alone contained fifty-three working collieries at one time, in an area only sixteen miles long. It was the most intensely mined area in the world and probably one of the most densely populated. From the rural population of around 951 in 1851 mass migration meant that by 1924 the population reached 169,000, approximately 20,000 people to the built-up square mile.

1855 is the date accepted as marking the change of the rural scene in Rhondda and the historic change to heavy industry, although coal was mined in the Rhondda as early as the seventeenth century for domestic purposes. The earliest date recorded for a safe coal level being opened in the Rhondda is 1790, mined by Dr Richard Griffiths, though it is disputable whether Dr Griffiths was the first to open a mine in the Rhondda. He is important as the first leaser of mineral rights in the area and as the builder of the first tram road into the Rhondda. The first real industrial pioneer in the area was Walter Coffin, who opened the first level and sunk the first pits in the Rhondda. With the discovery of the rich and prosperous steam coal seams of the Rhondda many more mines were to follow.

The Hafod concern was begun by two brothers, David and John Thomas. In 1850 they opened the Hafod Pit and although they reached the profitable Rhondda No.3 and Hafod seams, the workings proved unfruitful and were abandoned. The second venture took place at Coed Cae in 1850 by Edward Mills, again the workings were abandoned, this time owing to water seepage.

In the mid-1870s William Thomas Lewis, later Lord Merthyr, purchased the Hafod and Coed Cae shafts on the river Rhondda near Porth. The Coed Cae Pit was re-opened in the early 1870s and traded under the name of the Coed Cae Coal Co. The Coed Cae worked the upper bituminous (Household) seam coal only and closed in the 1930s. Hafod Pit is thought to have worked from the 1880s until 1893, working the bituminous seams, after which date the deeper steam coal seams were worked by Powell Duffryn.

By 1880 W.T. Lewis had sunk the Bertie shaft, and in 1890 the Trefor shaft (both Trefor and Bertie were named after W.T. Lewis' sons, and remain so today at the Rhondda Heritage Park). By 1890 the colliery was known as the Lewis Merthyr Navigation Collieries Ltd and from 1891 the five pits became the Lewis Merthyr Consolidated Collieries Ltd, employing some 5,000 men and producing almost a million tons of coal annually.

The Bertie shaft was 14ft 1in in diameter and 1,418ft 10in in depth. The winding engine was unique because of the unusual design of the drum known as a differential bi-cylindro conical drum, which enabled the engine to wind to and from different depths simultaneously. There is thought to have existed only one other engine of this style. The engine was originally steam operated until it was electrified in the late 1950s.

In 1904 the company sunk the Lady Lewis Colliery a mile to the north east in the Rhondda Fach and in 1905 they acquired the Universal Colliery at Senghenydd, which was later to suffer the worst ever mining disaster in British history. In 1929 the colliery became part of the Powell Duffryn Steam Coal Co., and in the same year Coed Cae stopped winding coal. Hafod No.2 followed, and Hafod No.1 in 1933. The colliery was nationalised on 1 January 1947.

The Rhondda No.1 coal seam was worked at the No.1 Rhondda Level, Trehafod, under the ownership of the Lewis Merthyr Navigation Collieries Ltd and from 1891 the level also known

Lewis Merthyr Colliery, The Rhondda Heritage Park and Trehafod in the New Millennium. The photograph was taken from the 'Tump' and was presented by Des Bird. The colliery height was 303ft 6in OD and was sited 670yds E. 23°, South of St Luke's Church, Porth. National Grid ref: 03969113. Sunk to 1,418ft 10in, Lewis Merthyr Colliery was formed from six shafts into the same workings.

as Rhondda Main Level came under the Lewis Merthyr Consolidated Collieries Ltd. In 1913 the level employed 234 miners and was closed in 1914.

In 1958 Lewis Merthyr Colliery and the neighbouring Tŷ Mawr Colliery merged and all coal winding ceased at Lewis Merthyr, with coaling continuing via Tŷ Mawr and men and supplies only at Lewis Merthyr. By 1969 the Colliery had become the Tŷ Mawr/Lewis Merthyr Colliery. As many as thirteen seams have been worked at the Lewis Merthyr using the advanced long wall method of working with most of the coal being won with pneumatic picks and hand loaded onto conveyors.

Although work was plentiful in the early years of the Rhondda industrial history, working conditions and pay were poor and disastrous. The cramped towns and bad sanitation led to ill health, poverty and death. Rhondda suffered excruciatingly hard and difficult times. Between 1868 and 1919 statistics show that a miner was killed every six hours and injured every two minutes. As a result of these conditions South Wales was at the forefront of political strife as the militant South Wales miners sought to ensure suitable working and living conditions in the Rhondda.

Until the 1950s the coal industry maintained a steady level of production and employment, but since that time there has been a continuing decline in the number of miners in employment. Most of the closed pits have still coal left to mine, but with oil and coal available more cheaply from abroad the demise of the industry has been inevitable. Nowhere has the decline of the coal industry been more dramatic than in the South Wales Coalfield.

Coaling ceased forever at Tŷ Mawr/Lewis Merthyr Colliery when the last dram of coal was raised on 21 June 1983.

By the end of 1990 not one productive colliery existed in the Rhondda but the spirit of the turbulent and proud Rhondda past has been captured and preserved as an historic landmark at the Lewis Merthyr Colliery, now the Rhondda Heritage Park.

W.A. Thomas in his later years. Sir William Thomas Lewis had sunk the Bertie shaft by 1880, and the Trefor shaft in 1890. By 1890 the colliery was known as the Lewis Merthyr Navigation Collieries Ltd, and from 1891 the five pits became the Lewis Merthyr Consolidated Collieries Ltd, employing some 5,000 men and producing almost a million tons of coal annually.

King George V (1910-1936) and Queen Mary at Lewis Merthyr Colliery on Thursday 27 June 1912. George, the second son of Edward VII and Alexandra of Denmark, was born at Marlborough House on 3 June 1865. Most of his childhood was spent at Sandringham, Buckingham Palace and Balmoral. After being educated at home by the Revd J.N. Dalton, George became a naval cadet at Dartmouth. By 1889 he was commander of a torpedo boat. However, in January 1892, his naval career came to an end when his older brother, Prince Edward, died of pneumonia. Edward had been engaged to marry his German cousin, Princess Mary of Teck. It was then decided she should marry George instead.

George was now heir to the throne and it was decided that he could no longer risk his life as a naval commander. He was granted the title the Duke of York and became a member of the House of Lords. George was also given a political education that included an in-depth study of the British Constitution. However, unlike his father, he did not learn to speak any foreign languages. George, Duke of York, married Princess Mary in 1893. Mary had six children: Edward (1894-1972); George (1895-1952); Mary (1897-1965); Henry (1900-1974); George (1902-1942); and John (1905-1919).

Edward VII died in 1910 during the Liberal Government's conflict with the Lords. His father had promised to give his support to the reform of the House of Lords if Herbert Asquith and the Liberal Party won a General Election on this issue. Although the 1910 General Election held in December did not produce a clear victory for the Liberals, George V agreed to keep his father's promise.

The outbreak of the First World War created problems for the royal family because of their German background. Owing to strong anti-German feeling in Britain, it was decided to change the name of the royal family from Saxe-Coburg-Gotha to Windsor. To stress his support for the British, the king made several visits to the Western Front. On one visit to France in 1915 he fell off his horse and broke his pelvis.

In 1924 George V, Monarch of the British Empire, appointed Ramsay MacDonald, Britain's first elected Labour Prime Minister. Two years later he played an important role in persuading the Conservative Government not to take an unduly aggressive attitude towards the unions during the General Strike. In an attempt to achieve national harmony during the economic crisis of 1931, the King persuaded MacDonald to lead a coalition government.

The following year, George V introduced the idea of broadcasting a Christmas message to the people. The King had not enjoyed good health for a long time and during his final years he spent much of his time on his grand passion, philately. Patriotically, he concentrated on collecting stamps from the British Empire. George V died of influenza on 20 January 1936. His eldest son, Edward became king.

South Wales Coalowners Association, August 1922. Some coal owners rose by their own efforts from fairly humble backgrounds to positions of importance. Many early coalowners were Welshmen, unlike the ironmasters who had nearly all come from England. They lived among their workers in the new mining communities of the valleys.

South Wales Institute of Engineers, 1923. Many coal owners did use some of their money to do good. They gave money for opening schools, chapels, workmen's institutes, libraries and hospitals. Many also became involved in charitable and social work, such as Lord Merthyr, William Thomas Lewis. Some irresponsible coal owners were called monstrous tyrant dragons, mistreating the entire workforce and communities to slavery and poverty.

Coed Cae Colliery (Housecoal Pit) in 1906. Sunk in 1850 by owners Edward Mills, it was reopened in the early 1870s and traded under the name of the Coed Cae Coal Co. Coed Cae upcast shaft was 9ft square, 117yds deep. Coed Cae downcast shaft was 16ft by 11ft ellipse, and 175yds deep. On 18 May 1872 the accident reports show that fifteen-year-old collier R. Thomas was killed by a fall of coal. In 1913 the manpower in the Coed Cae was 556. The Coed Cae Colliery ceased producing coal in 1929.

Coed Cae Colliery in the early 1900s. In 1893 a new horizontal twin-cylinder winding engine was installed. This engine was built by J.&W. Leigh, Ellesmore Foundry, Parricroft, near Manchester. Its cylinders were 26in diameter by 4ft 2in, 80lb psi steam pressure, the slide valves ran at 50rpm and producing 1,000ihp.

Coed Cae Colliery with its wooden headframe in the early 1900s. The winding engine was derelict by the 1960s and scrapped around 1973/1974. The site is now the Heritage Park Hotel and a monument has been erected dedicated to 'The Mining Communities of Rhondda who laid the foundations for the Valley's future prosperity'. Hwy Clod Na Golud, (Fame Outlasts Wealth). Unveiled by the Rt Hon. The Viscount Tonypandy of Rhondda on Sunday 6 October 1991 in the presence of The Mayor of Rhondda, Cllr Don May, and the Leader of Council Cllr, Mrs M.E. Collins MBE.

16

Coed Cae Colliery underground hauliers at work in the 1930s. The haulier's work was to drive the horse and empty dram to the coalface where the colliers would fill it with coal and then return to pit bottom with the full dram. He also had to feed and look after his horse during his working shift and, just like the miners, pit ponies had to have an annual holiday. Two weeks, come rain or shine, they kicked their heels as much as they liked in the green fields and pastures of Rhondda.

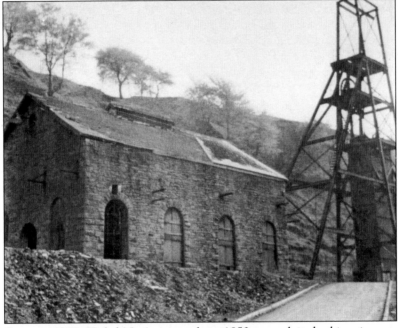

Hafod Colliery in 1938. Hafod No.1 was sunk in 1850 to exploit the bituminous seams, and after 1893 the deeper steam coal seams were worked. Hafod No.1 Pit was 16ft diameter and 461yds deep. Hafod No.2 Pit was 15ft by 9ft ellipse and 459yds deep. In 1909 Worsley Mesnes Ironworks Ltd, Wigan, received an order from Lewis Merthyr Consolidated Collieries Ltd for an engine for Hafod Pit. The order was for a horizontal twin-cylinder winding engine, with cylinders 34in diameter by 6ft stroke, 120psi steam pressure, a 15ft drum, drop valves at the top of the cylinders and Corliss valves at the bottom and producing about 2,500ihp. It was fitted with the tail rods. This engine was scrapped in the early 1970s. In 1913 the manpower in the Hafod was 1,130. In 1929 they were taken over by the Powell Duffryn Steam Coal Co., who remained owners until nationalisation in 1947. Hafod No.2 Pit ceased producing coal in 1930 and the Hafod No.1 Pit in 1933.

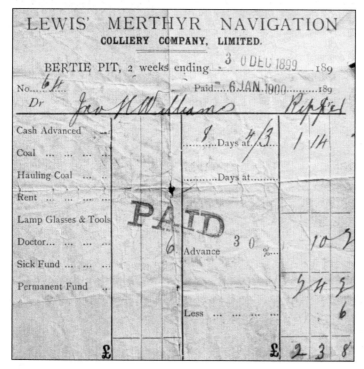

Left: Lewis Merthyr Consolidated Collieries Ltd, Hafod Pit Pay Docket, 1 February 1919. Labourer M. Edwards worked six days, and his take-home pay was £1 4s 3d (£1.21). Concessionary coal notes on the docket read: 'Only workmen who are householders are entitled to coal at reduced price and only for their own domestic use. Any other workmen wrongfully obtaining such coal will be prosecuted'. *Right*: Hafod-Rhondda Colliery Pay Docket, 23 September 1922. Thomas E. Kernin's take-home pay was 16s (80p).

Lewis Merthyr Navigation Colliery Co. Ltd, Bertie Pit Pay Docket, 30 December 1899. Ripper John Williams worked eight days underground at the Bertie Pit and received £2 3s 8d (£2.19). One old penny = 0.41 new pence, 12 old pennies or 1 shilling = 5 new pence, 240 old pennies or 20 shillings or £1 = 100 new pence or £1.

The Workmen's Compensation Act, 1906.

249   CLAIM FOR COMPENSATION.

I hereby give you notice that _William Rees_

of _138 Trehafod Rᵈ Trehafod_ .... claims from

you the sum of _35/—_ a week under and in accordance with the provisions

of the Workmen's Compensation Act, 1906, in respect of the injuries caused to him by

the accident which happened to him on the _Saturday the 22ⁿᵈ_ day

of _Decᵗ_ .... 19_23_, and in respect of which I served you

with notice of injury on the .... day of .... 191 ....

To ....

Dated the .... day of .... 192_3_

Signed _John J. Garwood_

Address _17 Nyth Bran, Llwyncelyn, Porth_

The Workmen's Compensation Act 1906 Claim for Compensation Receipt, 1923.
I hereby give you notice that William Rees of 138 Trehafod Road, Trehafod claims from you the sum of 35s a week under and in accordance with the provisions of the workmen's Compensation Act 1906, in respect of the injuries caused to him by the accident which happened to him on the Saturday the 22nd day of Dec 1923, and in respect of which I served you with notice of injury.
Signed John J. Garwood, 17 Nyth Brân, Llwyncelyn, Porth.

The Workmen's Compensation Act, 1906.

56   NOTICE OF INJURY.

Name of Person Injured _Isaac Roberts_

Address _38 Primrose Terrace Porth_

Occupation _(Trefor Pit) Checkweigher_

Cause of Injury _Strained ankle._

Date of Accident _18/11/25._   _Fell on Pit top._

As Agent on behalf of the above-named _I. Roberts_
I hereby give you notice that he has been injured whilst in your employment as above
stated.   To _L.M. Colliers_

Dated the _21_ day of _November_ 19_25_

Signed _John Treharne_

Address _55 Leslie Tree, Porth_

The Workmen's Compensation Act 1906 (1923) Notice of Injury Receipt.
Name of person injured, Isaac Roberts. Address, 38 Primrose Terrace, Porth. Occupation, Trefor Pit Checkweigher. Cause of injury sprained ankle. Date of accident 18 November 1925. As Agent on behalf of the above named I. Roberts I hereby give you notice that he has been injured whilst in your employment as above stated. To LM (Lewis Merthyr) Collieries dated the 21st day of November 1925. Signed John Treharne. Address 55 Leslie Terrace, Porth.

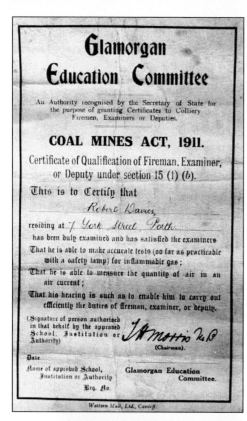

Glamorgan Education Committee

An Authority recognised by the Secretary of State for the purpose of granting Certificates to Colliery Firemen, Examiners or Deputies.

COAL MINES ACT, 1911.

Certificate of Qualification of Fireman, Examiner, or Deputy under section 15 (1) (*b*).

This is to Certify that

Robert Davies,

residing at *7 York Street Porth*

has been duly examined and has satisfied the examiners

That he is able to make accurate tests (so far as practicable with a safety lamp) for inflammable gas;

That he is able to measure the quantity of air in an air current;

That his hearing is such as to enable him to carry out efficiently the duties of fireman, examiner, or deputy.

(Signature of person authorised in that behalf by the approved School, Institution or Authority) *J H Morris M.P.*

(Chairman).

Date

Name of approved School, Institution or Authority   **Glamorgan Education Committee.**

Reg. No.

Western Mail, Ltd., Cardiff.

Glamorgan Education Committee Certificate, 6 December 1915. The Glamorgan Education Committee was an authority recognised by the Secretary of State for the purpose of granting certificates to colliery, firemen, examiners or deputies with the Coal Mines Act 1911 section 15 (1) (b), this one reads:

This is to certify that Robert Davies residing at 7 York Street Porth is duly examined and has satisfied the examiners that he is able to make accurate test (so far as practicable with a safety lamp) for inflammable gas, that he is able to measure the quantity of air in an air current, that his hearing is such as to enable him to carry out efficiently the duties of fireman, examiner or deputy. Deputies are appointed by the manager primarily to be in charge of an underground district and to have immediate charge of all the workmen in that district and of all operations carried out by them therein. Deputies are also required to carry out duties to secure the safety and health of all the men working there. The deputy is responsible, through the overman and undermanager, to the manager and must carry out their instructions.

Arthur James Cook 1884-1931. One memorable and leading mining figure to emerge out of the harsh political strife of the Rhondda was Arthur James Cook who resided at 52 Nyth Brân Terrace, Llwyncelyn, and started work as a labourer in the Trefor Pit. A.J. Cook soon worked his way to becoming a haulier. His political career was underway and he soon became a delegate to the Lewis Merthyr Employees Joint Committee while working at Coed Cae Pit. His political views were responsive to socialism and he became deeply involved in the South Wales Miners Federation in 1911.

A miners' open-air meeting in 1926. A.J. Cook later became the general secretary of the British Miners and led them during the struggles and hardship of the general strike in 1926. He led the miners by example, surrendering his union salary, living only on lockout pay and living on trains as he shuttled from coalfield to coalfield, preaching his revolutionary socialism to crowds of up to 100,000. Arthur James Cook led the fight for justice for the men of an industry in which a man died every six hours and another was badly injured every two minutes.

Left: Colliers Housecoal Curling Box in the early 1900s. The curling box was usually made in the blacksmith shop and was ready for use in a matter of minutes. *Right*: Rowland Morris on his first day at work in 1899. Rowland is seen here with his four-pint drinking water jack in the lining of his coat and food box ('tommy box') under his arm. He started work at thirteen years of age at Lewis Merthyr Colliery. Roland met King George V and Queen Mary on their visit to the colliery on Thursday 27 June 1912. Roland was killed by a roof fall while working at the Bertie Pit in 1951. In the 1880s and 1890s tinsmith Howel Thomas, formerly of Trehafod Road, hand made miners drinking water jacks and food boxes.

Gaffers and Ladies Pit Visit at Lewis Merthyr Colliery in the early years of the twentieth century. Girls and women were also employed below ground until the 1842 Mines Act. It is possible that women worked underground at a coalmine in the Rhondda Valley, but I have not seen any written evidence to confirm this possibility. Women worked underground as hauliers and doorkeepers. In 1913 the manpower in the Bertie Pit was 1,095, in 1947 the manpower was 1,107, and in the Trefor Pit 986. In 1954 with a combined manpower of 1,186, the pit produced 281,986 tons.

Lewis Merthyr Colliery in the 1950s. On the left is the Bertie Pit and on the right is the Trefor Pit with the newly constructed footbridge over the colliery sidings. In 1955, with a manpower of 1,506, it produced 249,211 tons; in 1957, with a manpower of 978, it produced 217,569 tons.

Lewis Merthyr Colliery Sidings in 1955. The 13-ton wagons of coal are ready and waiting for the brakesman. Their work included braking the full wagons of coal from the washery (Coal Preparation Plant) into the colliery sidings ready for delivery to the power stations, by-product plants and domestic users. Coal wagons varied in size and tonnage throughout the coalfield, but generally held 8, 10, 12, 16 and 21 tons.

Colliery Surface Worker with a dram, raced with 2 tons of coal in a 1-ton dram in the 1930s. The dram was designed to carry large lumps of coal, which was in great demand at the time. The collier cut the coal and kept his working place safe, working up to his knees in water and skilfully filling the drams with clean coal.

Lewis Merthyr Colliery Corn Stores in 1935. In 1930 pit ponies and horses consumed approximately 28,000 tons of corn and 30,000 tons of hay, a total of 58,000 tons of feed per annum. The animals were well cared for by the hauliers; stables generally were warm, dry and comfortable and many were lit by electricity. Moss litter or sawdust were always provided for bedding. Plenty of good food and clean water was at hand for the horses, both at the stables and while at work and a sufficient supply of medicine and dressing readily available.

Lewis Merthyr Colliery Pit Ponies in 1909. In 1930 there were approximately 11,500 horses employed underground in Coal Mines in Monmouthshire and South Wales. This represents approximately 4,197,000 horse days per annum, of which the maximum number of horse-working days would be 3,289,000, leaving 908,000 complete rest days per annum, or seventy-nine days per horse per annum. Several pit ponies had over eight years service underground and were treated like family pets by the hauliers.

Lewis Merthyr Colliery show pony Snowball in 1909. Behind Snowball is a Peckett Loco-motive and a wagon of pit props. In 1938 the Powell Duffryn Associated Collieries employed 1,450 horses. Coal Mines Acts extracts, General Regulations re Horses:

No horse shall be taken underground until it is four years old and until it has been tested by a duly qualified veterinary surgeon in the prescribed manner and certified to be free from glanders. All horses underground shall, when not at work, be housed in properly constructed stables, and in stalls of adequate size. No blind horse shall be worked in a mine.

Lewis Merthyr Colliery Peckett Locomotive in 1909. The Peckett was a very popular and useful colliery loco workhorse throughout the coalfield. It was used for supplying the screens with empty wagons and preparing the journeys of loaded coal wagons left behind by the brakesman ready for transportation to the power stations, by-product plants and the domestic coal users. The track was covered with frozen slurry in winter and a trip to the washery ended in slipping, almost stalling, and the last bit of the journey would then be completed at a struggling crawl.

Methane Drainage Pipe for Hafod No.2 Pit Shaft in the 1970s. The main constituent of firedamp is Methane (CH_4), which was produced during the formation of the coal measures. Therefore, it is contained in the coal seams and the associated strata and when they are disturbed by mining, the gas is released. It may come out slowly or escape as a blower. Methane itself has no smell, although firedamp may have an odour due to the presence of other gases mixed with it. High concentrations of firedamp can cause death by suffocation and this danger may be present in roof cavities, rise places, old working and unventilated headings. The gas is particularly dangerous because it can be set alight. The Mines and Quarries Act for deputies, 1954 (79) states:

If at any place in the general body of air 2% or more of flammable gas is found, you must, (a). Order your workmen to withdraw, (b). Notify as soon as possible your immediate superior and the person in charge of any other part of the mine likely to be affected by the presence of the gas, (c). Fence off all approaches to the affected area and mark with danger notices.

Colliery blacksmiths, fitters (mechanics) and ropesmiths in the 1950s. Their work at the pit was very important for the day to day efficient running of the colliery and included fully maintaining the surface layout and washery, the underground machinery and equipment, ropes, water and compressed air pipes, drams, pit guides, cages, winding ropes and ventilation fans, etc.

The Colliery Coke Ovens in 1910. On 14 May 1927 Rufus T. Lewis reported to the Lewis Merthyr Consolidated Collieries Ltd that he was pleased to say that output was improving:

The washery was very much out of date and it is not working at all satisfactory. You will remember that some time ago I brought up the necessity of having a spare 150hp motor for our Trehafod Collieries. Mr R.T. Rees took the matter up and prices were obtained but at the time we found nothing suitable. I certainly think we should have a spare. If something should happen to one of these motors we would have a serious problem.

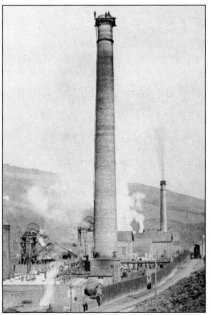

Left: Steeplejacks working on top of the chimneystack in the 1920s. In 1929, the collieries were taken over by Powell Duffryn Associated Collieries Ltd (also known as the 'Poverty and Dole' Group), who remained owners until nationalisation. *Right*: Lewis Merthyr Consolidated Collieries Ltd, Bertie Pit Nine-Feet Seam Pay Docket, 11 February 1921.

Bertie Winding Engine, Mining Department 1977, technical details:

No attempt shall be made by any person to drive this Winding Engine unless that person has been either:
(a). Authorised in writing by the Manager to drive this Winding Engine. (b). Authorised in writing by
the Manager to be trained to drive this Winding Engine under the supervision of a competent person.

General Data On Winding Apparatus. 14ft 0in to 24ft 0in Diameter BCC Drum. Drive: AC
geared with dynamic braking. Manufacturers: Electrical AEI, Mechanical Fowler (Robey Drum),
Brakes Dowler-Uskside – (Tredomen Brakes Posts). Date of Installation: Drum 1930, Brakes
and Electrics 1958. Motor Details: 900hp, 3,300v, 356rpm. Maximum Rope Speed: 34 ft/second
Men, 36ft/second Materials. Maximum Drum Speed: 27rpm Men, 28rpm Materials. Automatic
Contrivance: Lilly Duplex. Type of Brakes: Fowler-Uskside, LP Oil/Air Operated. Winding Levels:
Max 1,207ft. Intermediate Levels: Pump inset (Hafod Seam Pump House) 527ft, 4ft 0in inset (6ft
0in) 1,076ft. Shaft Conveyances: Two Cages – Single Deck – Two Drams. Type of Guides: Rope
– Bottom Weighted. Winding Rope: Two at 41mm Flattened Strand. Weight of Empty Cage and
Suspension Gear: 5.025tons. Maximum Number of Men per Cage: 26 Men.

Winding Engine Operation. Providing all the preceding requirements and checks have been carried
out and there are no other circumstances to preclude normal safe winding, then the drum may be
set into motion. Normal Winding Procedure. To ensure that the drum will move off in a controlled
manner, it is good practice to move the Power Control Lever to a position which by experience will
allow sufficient driving power in the correct direction before releasing the brakes. Men Winding. After
ensuring the drum moves off in a controlled manner as described above, the brakes should be moved to
the brakes 'OFF' position. Acceleration, maximum speed and deceleration will be achieved by Power
Control Lever and Brake Lever movements at the discretion of the Winding Engineman who will take
control manually throughout the wind. Maximum rope speed for men 34ft/second.

The Winding Engineman can select speeds lower than the normal cycle as necessary, by movement of
the Power and Brake Control Levers. The Winding Engineman must ensure brakes are in the fully 'OFF'
position when the braking function is not required and that whenever the drum has been brought to rest, the
Winding Engineman must ensure that the brakes are fully applied and that the Power Lever is in the 'OFF'
position. Before the Winding Engineman leaves the winding engine controlling gear he, should position the
cages in accordance with the colliery manager's instructions and ensure that the winding engine is immobilised.

Trefor Winder, Trefor Winding Apparatus Mining Department 1977, technical details:

No attempt shall be made by any person to drive this Winding Engine unless that person has been either: (a). Authorised in writing by the Manager to drive this Winding Engine. (b). Authorised in writing by the Manager to be trained to drive this Winding Engine under the supervision of a competent person.

General Data On Winding Apparatus. Type: Differential BCC Drum 10ft 0in to 17ft 0in; 14ft 0in to 22ft 0in. Drive: Two CA Cylinders. Manufactures: Electrical N/A, Mechanical Fowler, Brakes Fullerton Hodgart and Barclay (1976). Date of Installation: 1890. Motor Details: Two Cylinders 42in Bore, 72in Stroke. Maximum Rope Speed: 13ft/second – Men. Maximum Drum Speed: 11.3rpm – Men. Automatic Contrivance: Lilly Duplex. Type of Brakes: Fullerton Hodgart and Barclay. Weight/ CA Operated. Winding Levels: 1,377ft – Maximum Intermediate Levels. Shaft Conveyances: One Double Deck Cage – One Dram/Deck. One Single Deck Cage – One Dram. Type of Guides: Rope – Spring Tension. Winding Rope: Two at 41mm Flattened Strand. Weight of Empty Cage and Suspension Gear: 5.7tons and 4tons. Maximum Number of Men per Cage: 4 Men.

Winding Engine Operation. Providing all the preceding requirements and checks have been carried out and there are no other circumstances to preclude normal safe winding, then the drum may be set into motion. Normal Winding Procedure To ensure that the drum will move off in a controlled manner, it is good practice to move the Power Control Lever to a position, which by experience, will allow sufficient driving power in the correct direction before releasing the brakes. Men Winding. After ensuring the drum moves off in a controlled manner as described above, the brakes should be moved to the brakes 'OFF' position. Acceleration, maximum speed and deceleration will be achieved by Power Control Lever and Brake Lever movements at the discretion of the Winding Engineman who will take control manually throughout the wind. Maximum rope speed for men 13ft/second.

The Winding Engineman can select speeds lower than the normal cycle as necessary, by movement of the Power and Brake Control Levers. The Winding Engineman must ensure brakes are in the fully 'OFF' position when the braking function is not required and that whenever the drum has been brought to rest, the Winding Engineman must ensure that the brakes are fully applied and that the Power Lever is in the 'OFF' position. Before the Winding Engineman leaves the winding engine controlling gear he should position the cages in accordance with the colliery manager's instructions and ensure that the winding engine is immobilised.

General View of Lewis Merthyr Colliery in the early 1900s. In 1891 the Lewis Merthyr Consolidated Collieries Ltd acquired the Llwyncelyn Colliery, Porth, and can be seen in the centre of the photograph. Sir William Thomas Lewis always gave the opportunities and assistance to mining students and members of the mining college who visited the colliery for inspection during their mining education, which also included First Aid in the coalmines and for which the Priory for Wales, Order of St John, was mainly responsible. Deputies had first aid qualifications and it was desirable that there was one first aid man for every thirty workmen. First Aid containers and morphia safes were near to concentrations of men and regularly moved up near to the face.

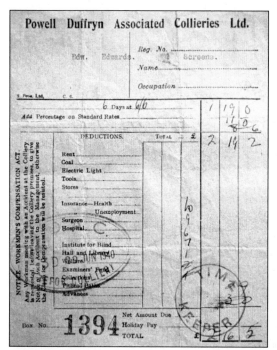
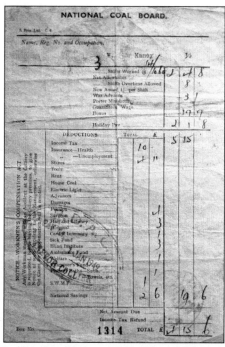

Left: Powell Duffryn Associated Collieries Ltd (PDs) Pay Docket, 29 June 1940. Edward Edwards received £2 16s 5d for six days work in the colliery screens (coal preparation plant). The PDs owned the colliery prior to nationalisation on 1 January 1947. *Right:* NCB (National Coal Board) Pay Docket 1 January 1949. W. Kaney's take home pay was £4 15s 6d.

Lewis Merthyr Colliery picking belt in the 1950s. Their job was to inspect the coal, which was tipped from the drams via the tipplers, onto moving belts, picking out any stone, timber, steel or any other rubbish. The surface workers were usually ex-colliers who through injury or ill health were no longer fit enough to work below ground. Their job was very hard and the conditions very dusty and noisy. They were the lowest paid of all miners.

Lewis Merthyr surface and incline workers in the 1950s. The photograph includes: Jack Thomas, Ronnie Downs, Johnny Jackson. The men who worked on the incline and the slag heaps (tips) worked in blizzards and Arctic conditions during the winter months. The rails often froze over causing journeys of colliery waste to be derailed.

Hancocks Beer Advertisement, 1953.

'Black diamonds' – the wealth of Wales. 23,687,500 tons of coal were mined in South Wales and Monmouth during 1953. After a hard day at the coal face-just hand me a Hancocks. If Hancocks brew it, be it draught or bottled, you'll prefer a Hancocks every time. Here is beer with strength and character, a drink that puts you back on your feet-a very fitting reward for a hard day at the 'face'. If you want beer as it should be, insist on Hancocks. Hancocks keep alive all the fine traditions of brewing, providing a beer that satisfies and refreshes; a beer that is all the national drink should be. Take a couple of Five-Fives home with you.

Hayden Nilos Conveyor Belt Fastener advertisement.

Hayden-Nilos machines make perfect joints with minimum effort. Used with Hayden-Nilos precision ground taper-pointed hooks they ensure trouble free belt joints and are firm favourites in the coalfields. No belt holdups to spoil your record efforts with Hayden-Nilos joints; many of them have been known to run three years or longer without a break. In a large Midland colliery, on an important and exceptionally heavy-duty unit, they renew all joints every four months and report that this regular precaution ensures trouble-free conveying and pays handsome dividends in coal production.

Explosion At Lewis Merthyr Colliery, Glamorganshire

Report on the causes of, and circumstances attending, the Explosion, which occurred at Lewis Merthyr Colliery,Glamorganshire, on 22nd November, 1956

by

T.A. Jones, OBE HM Divisional Inspector of mines.

Presented to parliament by the minister of Power by Command of Her Majesty December 1957.

The Right Honourable Lord Mills, KBE   8th November 1957.
Minister of Power.

MY LORD,
INTRODUCTORY

In accordance with your direction, given under the terms of Section 121 of the Mine and Quarries Act, 1954, I beg to submit my report on the causes of, and circumstances attending, the explosion which occurred at Lewis Merthyr Colliery, Glamorganshire, on 22nd November, 1956, when gas in a roadhead cavity was ignited. As a result of the explosion two persons were killed and seven others died of their injuries. Five other persons were injured. A list of the casualties is given in the Appendix.

On 1st March, 1957, Mr T. Alwyn John, HM Coroner for North Glamorgan, sitting with a jury, concluded the holding of an inquest on the bodies of the nine deceased persons. A verdict was recorded that all nine had "died as a result of burns accidentally received in an explosion at the roadhead of the centre road in the "N4" district of the Two-Feet-Nine Seam at Lewis Merthyr Colliery on 22nd November, 1956".

I. DESCRIPTION OF THE COLLIERY

General

The Lewis Merthyr Colliery is situated in the village of Trehafod some 18 miles north of Cardiff. The present main winding shafts for men and mineral, known as Bertie Pit and Trefor Pit, were sunk in 1878 but high-class steam coal had been produced some ten years earlier from shafts still contained within the Lewis Merthyr mine. There are, in all, six vertical shafts serving the mine. Of these, the House Coal shaft and the Cymmer shaft are used only for pumping purposes. The Lady Lewis shaft is used only for ventilation, serving as an upcast shaft. The Hafod shaft is used as a downcast shaft and also serves as a third means of egress to the surface.

The output of the colliery is some 1,250tons of coal per day. The number of persons employed underground is 936, and 226 are employed on the surface.

Management

The mine is in the No.3 Area of the South Western Division of the National Coal Board. The principal officials were:

Mr M.J. Davies, Area General Manager.
Mr A. Hudson, Assistant Area General Manager.
Mr G. Tompkin, Area Production Manager.
Mr C.H. Hodkin, Deputy Area Production Manager (Operations).
Mr D.N. Simpson, Deputy Area Production Manager (Planning).
Mr J. Murphy, Group Manager.
Mr A.R. Fox (fatally injured in the explosion), Manager.

Mr E. Moore, Under-manager for Trefor Pit.
Mr W. McDonald, Under-manager for Bertie Pit.

Seams Worked and Lamps Used
The names of the seams being worked, given in descending order, are Two-Feet-Nine, Six-Feet, and Gellideg. All the seams are worked by the advancing longwall method, the coal being won with pneumatic picks and hand loaded on to conveyors. The only seam affected by the explosion was the Two-Feet-Nine seam where the depth of cover was 328yds.

Safety lamps have always been required to be used throughout the mine. The electric lamps used by the workmen are Oldham type GW cap lamps. To comply with the Regulations relating to Firedamp Detectors, Cambrian type No.8 flame lamps are used. For inspection purposes Prestwich Patent Protector No.6 flame lamps are used.

The Two-Feet-Nine Seam
The seam is the Four Feet in Woodland's correlation. It has an aggregate thickness of seven feet and contains four dirt bands of varying thickness. The roof is a good Clift and the floor is fireclay. The volatile content on an ash free dry basis is 18.5 per cent. The seam has been worked fairly extensively in the locality.

II. THE 'N4' DISTRICT

Method of Work and Organisation
At the time of the explosion the 'N4' district was the only district being worked in the Two-Feet-Nine seam. The coal produced was raised at the Trefor pit and comprised the total output of that pit, about 320tons per day.

This district consisted of a double unit longwall conveyor face served by two end supply roads and a centre loading road. The gradient was negligible. Each side of the face was 105yds long. The coal was hand got with the aid of pneumatic picks, and the face was advanced four feet six inches on a 48-hour cycle. The coal was filled on the morning shift only, with the cycle of operations arranged so that the face conveyors were moved forward on alternate afternoons. There was a similar alternation in the waste drawing and packing operations done on the afternoon shift. The three roads were ripped on afternoon and night shifts, the rippings being taken entirely in the roof.

Traditionally the working section in this seam had been the lower four feet of the seam with a bed of coal one foot six inches thick forming the roof, and the 'N4' panel had been opened in this way. In September, 1956, it had been decided to try to work the full thickness of the seam. By so doing coal would not be lost and better material would be obtained for the packs. By November the whole of the left hand face was being worked in the full thickness of the seam, but the right hand face remained as originally developed.

All machinery was driven by compressed air. No electrical power was installed in the district.

On each of the three shifts there were two deputies. One deputy was responsible for the working face, for the end roads to the junctions with the intake airway and for the centre road to its junction with the return airway. The other deputy was responsible for the intake and return roads outside these junctions. An overman was in charge of the morning shift; the night shift was similarly supervised.

Support and Control of Roof
The support on the left hand face was by Dowty hydraulic props set to heavy section corrugated steel bars. The bars were set at three-feet intervals with rows of props four feet six inches apart.

Strip packs were built eight yards wide leaving wastes ten yards wide. In each waste Walton quick release steel chocks were set along the waste edge, from six to nine of these chocks being set in each waste. On the right hand face-where the traditional working section was taken-the roof was carried on H section steel props, having both ends closed, set to corrugated steel bars. Wood compression pads were used between props and bars. The rows of props were set at four feet six inches apart the bars four feet apart. On this face, strip packs five yards wide were built leaving wastes six yards wide. Opposite each waste two hard wood chocks fitted with quick release devices were built.

The coalface at the roadhead of the centre road was kept about 12ft advance of the general line of face, and the rippings were taken to the coal head. The roof was supported by means of steel arches, 14ft wide, set at intervals of four feet six inches to correspond with the advance of the face conveyor. These steel arches were a temporary form of support for the span of the roof, about 20ft long, between the coal head and the permanent roadway supports. They were replaced immediately outbye the face conveyors by the permanent supports, which consisted of steel arches of similar dimensions but having straight portion of each leg cut off and a curved steel plate welded on, which rested on soft wood cogs forming part of the ten yards wide roadside packs. These permanent supports were set three feet apart, the cogs being erected the height of the face working.

The end supply roadways were supported in the same way but the interval between the inbye permanent support and the face did not exceed ten feet as these roadways were not advanced in front of the general line of face.

Ventilation and Occurrence of Firedamp
This district was ventilated by a current of fresh intake air from the Hafod downcast shaft, which returned direct to the Trefor upcast shaft.

The end supply roads served as intakes, with the centre loading road serving as a common return. Air measurements made on 5th November 1956, showed 7,828cu.ft of air per minute entering the left hand face, 7,201cu.ft entering the right hand face, and 15,442cu.ft returning on the centre road. Air samples taken at this time showed that the methane content in the return air in the centre road was 0.60 per cent. The highest methane content recorded on this road was 0.80 per cent, on 12th June, 1956, when the quantity of air passing was 11,687cu.ft per minute.

The reports made by the deputies showed that inflammable gas had been found on four occasions since the district started production in February, 1956. All reports were of 'blowers diluted at point of issue' at the waste edge.

Use of Explosives
Explosives were not normally used in this district. The roof rippings in all three gate roads were got down by means of pneumatic picks. The only case of a shot being fired anywhere in the district was on 16th November, 1956, when one shot was fired in the floor at the centre road. The shot was necessary to grade the road through a fault.

Precautions against Coal Dust
For the suppression of airborne dust, the seam was systematically water infused with satisfactory results. The roadways were regularly treated with limestone dust. The most recent roadway dust samples taken before the explosion in the centre road showed the incombustible content of the dust on the floor to be 78 per cent, and on the roof and sides 75 per cent.

III. EVENTS PRIOR TO THE EXPLOSION

On the night of 8/9th November, 1956, an extensive fall of roof occurred in the roadhead of the centre road from the inbye permanent support practically to the face of the roadhead, a length of some 16ft. One steel arch was left standing between the inbye end of the fall and the coal face. The cavity was the full width of the roadway and exposed the Three Coals seam some 24ft above. For some time earlier a small fault had been working down the left hand face towards this road. At the time of the fall this fault was less than ten yards from the left hand side of the road. There was no evidence of this fault in the cavity, but it was obvious that the thick bed of clift above the seam had changed to become weaker than normal and lacking in its usual cohesion. The fall was cleared and 14-feet steel arches were erected, beneath the cavity. These arches were covered with wood lagging which in turn was covered with a 'cushion' of rubbish some four feet thick, the top of which would thus be some eight feet from the top of the cavity. The roof and sides of the cavity above this packing were not supported in any way. The production of coal was resumed on Monday, 12th November, 1956.

Work proceeded without untoward incident until the night of 19/20th November, 1956, when a second fall occurred at the roadhead. This was an extension of the earlier fall. The cavity now extended to the coal head and was some 30ft long and 30ft high. It had also widened to about 30ft exposing a slicken-sided slant some ten feet to the left of the fault previously mentioned, which was now crossing the middle of the centre road. There had been no earlier indication of the presence of this slant. This second fall made coal production impossible and this situation was unchanged on 21st November. By the afternoon shift of this day, the fall had been cleared and the erection of steel arches beneath the cavity was begun. This work was being carried on by the night shift when, at about 3:00a.m. on 22nd November, four of the six newly erected steel arches were displaced by a stone weighing about three tons which fell from the cavity. The colliery manager, accompanied by the morning shift overman, arrived at the scene at about 5:30 a.m. He decided to erect an 'umbrella' of ten-feet arches covered by wood lagging beneath which the gate conveyor could run. These ten-feet arches could be erected without disturbing the 14-feet arches displaced by the fall. The stone which had fallen was broken up by means of a pneumatic pick and the work of erecting the steel arches begun. By this time the men employed on the morning shift had begun to reach the meeting station at the junction of the left hand supply road with the intake airway. A few of these men were brought forward to assist with the work and the remainder told to stay at the meeting station until they received further instructions.

IV. NARRATIVE OF THE EXPLOSION

By about 7:15 a.m. three of the ten-feet arches had been erected. Fourteen persons were variously engaged in the work. Two workmen were standing on a staging, tightening the fishplate bolts, and four others were holding the legs of the arches. The night shift deputy had gone back in the road some 30yds to a point where a repairer was preparing wood struts for use between the arches. The others were standing, prepared to cover the arches with wood lagging, when a further fall occurred from the cavity. The fall was of some two tons of stone, most of it in one piece. Almost coincident with the fall there was a flame. One of the workmen on the staging stated in evidence that he heard the fall and jumped from the staging. As he jumped he saw the flame. The deputy heard the fall and looked inbye. He stated that fall and flame were simultaneous.

All the persons present were enveloped in flames and suffered severe burns. Two died from their burns at the scene and seven others died later in hospital.

The men at the meeting station noticed a 'puff' of wind and a cloud of dust. A collier who had passed into the left hand face returned to say that he had seen flame in the centre road. The work of rescue was quickly organised. The first aid and rescue work will be dealt with in a later section of this report.

V. THE CAUSE OF THE EXPLOSION

The Nature of the Explosion
From all the evidence it was obvious that an explosion of firedamp had occurred in and

beneath the cavity. Flame had been projected some 70yds outbye along the centre road, for about 15yds along the left hand face and 25yds along the right hand face. The severe nature of the burn injuries sustained by many of the casualties, suggested that flame had persisted for an appreciable time in the vicinity of the cavity. There was no sign of violence and no indication that coal dust had played any part in the explosion.

The Source of the Firedamp

Tests made during the investigation showed an explosive mixture of firedamp and air nine feet down from the top of the cavity; the methane content at the top exceeded 80 per cent. The normal emission of firedamp from the seam, probably augmented by emission from the Three Coals Seam exposed by the fall, would naturally produce these conditions in an unventilated cavity such as this and it can safely be assumed that similar conditions obtained immediately before the explosion, although the presence of inflammable gas in the cavity had not been detected.

Immediately after the fall on 20th November the under-manager and the deputy climbed up on the debris, examined to the top of the cavity and found it clear. At this time the heap of fallen debris was, of course, deflecting at least part of the air current into the cavity. As the fall was cleared the top of the cavity became increasingly inaccessible both to the air current and for examination. No steps were taken to direct an air current to the upper part of it or to enable examinations to be made there. When the fall was finally cleared the cavity was entirely unventilated and could not be examined as no means of access had been provided. About two hours before the explosion, the night shift deputy stood on the tops of the steel arches and tested for gas at the highest point he could reach, but this was about 15ft above the floor and 18ft from the top of the cavity.

The Igniting Medium

All possible means of ignition were carefully investigated and considered.

The electric and flame safety lamps which were in the district at the time of the explosion were sent, in the condition in which they were brought out of the mine, for examination and testing at the Safety in Mines Research Establishment. None of the flame lamps exhibited any defect likely to constitute a hazard. All the electric lamps showed signs of heating and in three cases the heat had been so intense that the cable sheathing and the core insulation had been burned away so as to expose the bare conductors. This damage had obviously been caused by the flame of the explosion. Only one of these electric lamps showed signs of damage other than that caused by heat. In this case the headpiece had been broken so as to expose the main and pilot bulbs, both of which were broken. The filament of the pilot bulb was intact and was bright and clean, showing that it had not been heated in air. The whole of the coiled centre portion of the filament of the main bulb was missing, only small portions of the straight part of the filament remaining attached to the filament supports, which had been bent over sideways to an almost horizontal position. The filament supports were nearly touching, but there were no signs of arcing when examined under a microscope. Microscopic examination also showed that the filament ends, where broken, were angular rather than rounded, suggesting breakage and not fusion. The only occurrence at or about the time of the explosion which might have caused this damage was the fall, and investigation revealed that the person using this lamp was standing at least five yards outbye from the point where the stone fell. The damage could easily have been caused in the confusion following the explosion.

On this evidence it was concluded that none of the safety lamps was the igniting medium.

After the explosion, the compressed air hose from a manifold on the centre road to the turbine of the right hand face conveyor was found to be leaking badly from a hole. There was also a slight leak of compressed air from a joint in the two-inch pipe range laid along the floor. These items of equipment were cut out and sent to the Safety in Mines Research

Establishment for examination. In both cases the damage was found to be post explosion, and they were dismissed as possible means of ignition.

A compressed air lamp was found lying in the roadside some 30yds outbye the cavity. This lamp had been used at the face conveyor transfer point during coaling operations, but there was conclusive evidence that it had not been in use since the second fall occurred.

Except for two mining type telephones, electricity was not used in the district. These telephones were found to comply fully with their certification specification and standard of safety when tested at the Safety in Mines Research Establishment. Later evidence disclosed that these telephones had not been in circuit for some days before the explosion.

Nothing was found in the investigation to suggest that the use of any article of contraband was responsible for the explosion.

The only possible source of frictional heating lay in the belt conveyors. There was definite evidence that none of the conveyors had run during the shift.

It will be recalled that the statements of survivors left no doubt that the explosion was coincident with the fall of roof. This led to a careful investigation of the possibility of incendive sparks having been produced by the fall. The stone which fell was estimated to weigh nearly two tons and was composed of hard clift containing ironstone. The only ground of this nature was near the top of the cavity. The stone had therefore fallen fully 20ft before reaching the arches. Samples of the stone which fell were sent to the Safety in Mines Research Establishment and were subjected to a variety of tests. These tests did not produce a conclusive result but, after consideration of all the factors involved, including an appreciation of ignitions previously obtained experimentally using similar stones, the opinion was formed that incendive sparks could have been produced by the impact of a piece of the hard clift from the top of the cavity on one of the steel arches about 20ft below. Although no trace of inflammable gas was found before or after the explosion at the horizon where the stone struck the arches, tests carried out at the Safety in Mines Research Establishment showed that, under certain conditions, an object falling from near the top of a cavity containing a high concentration of methane could bring down enough methane to produce an explosive mixture at the base of the cavity.

Careful consideration of all the available evidence has led me to form the following conclusions:

1. After the fall had been cleared and while the steel arches were being erected, the upper part of the cavity contained a high concentration of methane.
2. The stone falling from near the top of the cavity brought down enough methane to produce an explosive mixture at the horizon of the steel arches.
3. The impact of the stone striking the steel arches produced an incendive spark, which ignited the explosive mixture, whereupon flame spread to the extent described earlier in this report and persisted until all the methane had been consumed.

V1. MATERS ARISING FROM THE EXPLOSION

Support of the Centre Roadhead
The fall which occurred on the night of 8/9th November showed clearly that the system of support, by steel arches four feet six inches apart, at the roadhead between the permanent roadway supports and the face was not effective in controlling movement of the strata and keeping the roadhead secure. The presence of the fault undoubtedly contributed to the deterioration in the strength and cohesion of the roof strata disclosed by the fall. This deterioration was probably the main reason why a system of support which had hitherto proved effective now failed completely, but the collapse of four adjacent steel arches was evidence that they could not have been properly strutted and tied together so as to provide stability.

It should have been obvious to the management that this system of support could no longer be relied on and that improvements were vitally necessary, yet, when the face restarted after the fall had been cleared, the same system was adopted without any modification. The second fall which occurred on 19/20th November was the natural sequel.

The Fall Cavities
In dealing with each of the two falls there was a complete disregard for the requirement of Regulation 8 (b) of the Coal Mines (Support of Roof and Sides) General Regulations, 1947, in that no attempt was made to set supports to the newly-exposed roof and sides.

The setting of some form of support to the newly-exposed roof and sides should have been commenced while the heap of debris provided access to the top of the cavity and should have been continued as opportunity occurred during the removal of the debris. As soon as the stage had been reached where fresh roadway supports could be set, the cavity should have been stowed solid above the supports or, if this were impracticable, the roof and sides of the cavity should have been secured by means of cogging and strutting above the steel arches, and means should have been provided for ventilation of the cavity and for access to the top of it so that it could be properly examined on each deputy's inspection.

The 'umbrella' form of support affords no real protection to persons working or passing in the roadway, and merely engenders a false sense of security. This occurrence, and others of a similar nature, notably that at Cwm Colliery discussed in the Divisional Inspector's Annual Report for 1949, emphasise its potential, even if remote, dangers and the effect is alarming when the prevalence of such cavities is considered together with the probability of firedamp accumulating at the top of them. The falls of ground which cause these cavities can largely be eliminated by the adoption of a system of roof control founded on a proper appreciation of all the relevant details in any particular conditions. No cavity should be allowed to remain unventilated.

V11. FIRST AID AND SUBSEQUENT TREATMENT

The first alarm was received by the morning shift men who were waiting at the meeting station to be passed into the district. Amongst these men were two very capable first-aid men and they, with the deputies, took charge of the situation. A good supply of first-aid material was available in the ambulance station at the entrance to the district, and these supplies were augmented from a second ambulance station some 500yds outbye. The colliery emergency organisation was brought into operation at once, and a rescue brigade from the nearby Porth Rescue Station arrived very promptly. The recently installed system of wireless communication between rescue stations and rescue vans was used to reinforce this brigade at the pit. This was the first occasion of its use in actual emergency in this Division.

The casualties received first-aid treatment without delay and were conveyed to the Medical Centre by way of the Hafod shaft which offered the quickest route. At the centre they received further treatment from doctors, the sister in charge and first-aid men before being conveyed in motor ambulances to hospital. Two of the men died at the scene of the explosion and two others died within 12 hours at East Glamorgan hospital. One was detained at this hospital and recovered. The remaining nine casualties were transferred to the Chepstow Burns Unit, where five of them died later.

VIII. ACKNOWLEDGMENTS

I wish to record my grateful appreciation of the helpful co-operation of all who took part in the investigation of this occurrence on behalf of the National Coal Board, the National Union of Mineworkers, the British Association of Colliery Management, and the National Association of Colliery Overmen, Deputies and Shot Firers; and to express my thanks to the officers of the Safety in Mines Research Establishment who took part in the investigation and carried out the tests, and to Mr T. Hillman, Area Surveyor, National Coal Board, who prepared the plan which accompanies this report.

I have the honour to be, my Lord,
Your Lordship's servant, T.A. Jones

LIST OF CASUALTIES
Killed

Name	Age	Occupation
E. Howells	37	Turbine attendant
S. Thomas	69	Overman

Fatally Injured

A. Atkins	40	Collier
T. Davies	38	Chargeman
A.R. Fox	41	Manager
C. Jones	36	Overman
R. Jones	57	Repairer
J.H. Mills	35	Repairer
P. Profitt	27	Repairer

Injured

H. Bryant	46	Chargeman
W. Childs	43	Repairer
F. Crump	39	Repairer
W.H. Davies	50	Packer
I. Humphries	35	Deputy

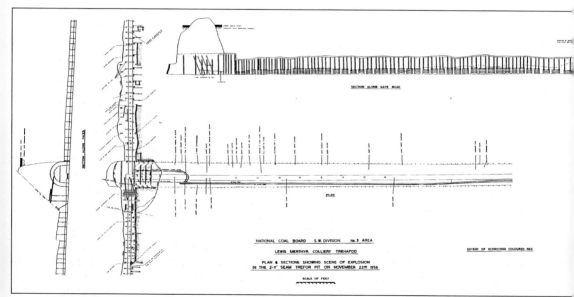

Plan of Explosion

Mining Fatalities: the Price of Coal

Life underground, down in the bowels of the earth, was hazardous but coalmining was a daily and regular feature for me all my working life.

If asked to describe life underground I would say, 'It's not easy with the appalling conditions, roof breaking, timber creaking, stones falling, coal and stone dust rising, water dripping, poor ventilation, and in the early days of mining using naked flame lamps and inadequate safety flame lamps giving insufficient light (about one candle light), sweat running, blood seething and, on some occasions, breath failing.' Yet this was the way we earned our bread and butter.

The number of hours worked underground was twelve hours per day at the end of the nineteenth century; this was then reduced to ten hours per day. The objective was for a reduction to eight hours per day. A jingle was written: 'Eight hours work, eight hours play, eight hours sleep, and eight bob a day.'

Records show that hundreds of young coalminers had to cease work due to the dreaded disease of silicosis, which was considered a progressive disease due to the lungs hardening like cement. It caused painful suffering and often culminated in loss of life, very often before the sufferer had reached his twentieth birthday, indeed a very young age.

Many families were deprived of the breadwinner. It was easy to understand why so many fathers made many sacrifices and much effort to keep their sons away from the Pits, and to follow a safer calling and occupation.

There were many accidents at the Pits, on the surface and underground. There were many explosions and fatalities were high, especially in the early years of coalmining.

Today, many ex-miners are still suffering with pneumoconiosis, emphysema, bronchitis, vibration, white finger, etc. and are still waiting for compensation from British Coal.

Lewis Merthyr Colliery and walkway from the pithead baths and leading to the colliery lamp room in 1972. The miners cap lamp and naked flame safety oil lamp were kept in the colliery lamp room. At the commencement of each shift, every underground miner would report to the lamp room, take his cap lamp and place a check on a hook above its place indicating that he was in work. He would then report to the Lodge – the name given to the office where officials gave their orders, and get his instructions. From there he would make his way to the top of the pit for his 'bond'. A bond is a name given to the pit cage when carrying men through the pit shaft. An official of the mine would check his lamps for safety.

Lewis Colliery walkway and lamp room in 1974. District names in 1927 at the Bertie Pit: Clapps (9ft), Old Red Coal (6ft), Everleighs (Red Vein), Plummers (Red Vein), Main North (6ft), Henry Williams' (6ft), Six Feet Main (6ft). Trefor Pit: Llantwits (Lower 5ft), Far End (Lower 5ft), Knox's, Brown' s (Upper 5ft), Cook's (Upper 5ft), Hard Heading (Lower 5ft). Hafod No.1 Pit: Wymans (Middle 5ft), West 'John Williams' (Upper 5ft), No.1 Slant (Upper 5ft), No.2 slant (Middle 5ft), Ladysmith (Upper 5ft), Woodlands (Upper 5ft). Hafod No.2 Pit: South (Working, Coed Cae and Eirw property), South East (Working, Hafod Fawr, Gellywion, Coed Cae and Eirw properties), West (Working the Llwyncelyn, Coed Cae and Eirw properties, North East (Working, Hafod Ganol property) and the North West District.

Left: Lewis Colliery Lamp room in the 1950s. *Above* Lewis Merthyr Colliery, NCB lamp check. At the end of his shift the miner would replace his cap lamp to be charged, and his oil lamp to be cleaned, refuelled and checked for safety. He would then remove his lamp check showing that he had arrived safely to the surface from the pit. In the event of an explosion it would instantly show who was still underground.

Lewis Merthyr miners ready to descend the pit in 1906. On top of the pit at the pit bank there were three tipplers (tumblers). The tipplers would invert and tip the drams of coal onto a moving conveyor directly underneath and convey the minerals into the washery (coal preparation plant) for cleaning, washing and grading the coal.

Bertie Pit Bank (top of the pit), 1983.

At every mine where persons are carried through a shaft the manager shall make and secure the efficient carrying out of arrangements, whereby a competent person appointed by him for the purpose called a 'banksman' is in attendance for the purpose of receiving and transmitting signals at the landing (Pit Bank) in use at the top of the shaft: (a). Whenever any person is about to be lowered through that shaft; and (b). Whenever any person who is to be raised through that shaft is below ground.

Bertie Pit Bottom, 1906.

At every mine where persons are carried through a shaft the manager shall make and secure the efficient carrying out of arrangements whereby whenever any person who is to be raised through that shaft is below ground, a competent person appointed by him for the purpose as an 'onsetter' (Hitcher) is in attendance for the purpose of receiving and transmitting signals at the entrance to that shaft from which any such person is to be raised.

Bertie Pit Bottom, 1983.

Procedure when persons are to be raised. (1). No onsetter or other person authorised to transmit signals shall allow any person who is to be raised to the top of a shaft to enter a cage for that purpose, (a). Unless he has transmitted to the banksman the signal 3; and (b). Unless he has received from the banksman the signal 3. (2). In order to direct the person operating the winding apparatus to raise a cage when any person is therein the onsetter or other person authorised to transmit signals shall transmit to him the signal 1. (3). The person operating the winding apparatus shall not raise a cage in pursuance of a signal 1 given under the last preceding paragraph until he has received from the banksman the signal 2. Signal to stop cage in motion when persons are carried, 34. When the cage is in motion and any person is therein the person operating the winding apparatus shall stop the cage upon receiving the signal 1. Signals when persons are not carried, 35. In a shaft when persons are not being carried the following signals and no other shall be transmitted to require the movements specified in relation thereto, that is to say to raise up 1, to lower down 2, to stop when in motion 1, to raise steadily 4, to lower steadily 5.

On the left of the photograph the colliers are hand cutting the coal. On the right, a collier is making safe the roof, which is of good Clift and the floor is fireclay, which can be used for brick making. Naked flame lights are in use as it was the early 1900s. Throughout the ten years 1851-1860, no less than one-third of the principal disasters in coalmines throughout Britain occurred in the South Wales Coalfield, though Britain's coal output was nearly ten times as great during that period.

Having a food break ('grubo') in the early 1900s. Naked flame lights are in use; flat caps (Dai Caps) were used before safety helmets, flagon bottles were used for drinking water and metal tommy boxes were used to keep the mice or rats getting at their food. From 1850 to 1920 no less than 3,179 colliers of South Wales were killed in disasters in the region compared with 8,520 for Britain as a whole, and of twenty-seven disasters in UK, from 1890 to 1913, thirteen occurred in South Wales. Rhondda's tally was especially heavy, though under merciful providence no explosion caused as much loss of life as at the Universal Colliery, Senghenydd (439) on 14 October 1913.

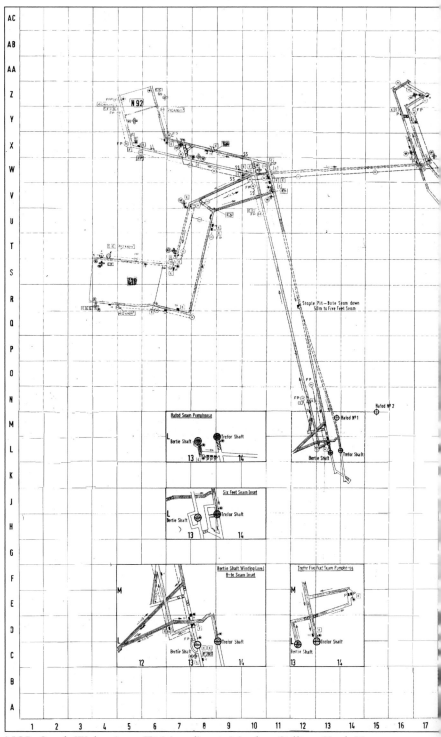

NCB, South Wales Area, Tŷ Mawr/Lewis Merthyr Colliery Yard, Upper Nine-Feet and Four-Feet Seams, Firefighting and Rescue Plan Section, 26 April 1983.

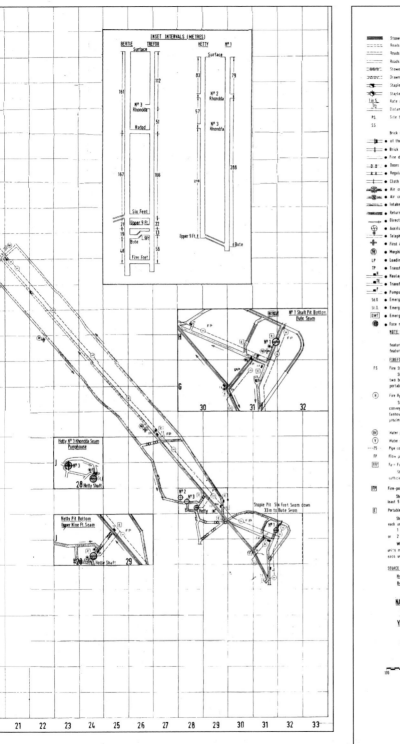

REFERENCE

Symbol	Description
	Stone drifts
	Roads in goaf
	Roads in goaf one fast side
	Roads in solid
	Stowed roads
	Drawn off roads
	Staple shafts
	Staple shafts abandoned
1 in 5	Rate and direction of dip
	Distance from shaft in kilometres
PS	Site for emergency stopping (prepared)
SS	(selected)

Brick stone or concrete stopping constructed as described in Regulation 71(2)
● of the Ventilation Regulations 1956
● Brick stone or concrete stopping other than above
● Fire dams or seals
● Doors
● Regulators
● Cloth
● Air crossing (explosion proof)
● Air crossing other than above
● Intake airways
● Return airways
● Direction of air current red or blue arrows
● Auxiliary fan
● Telephones
● First Aid stations in red
● Morphia stations in green
LP ● Loading points
TP ● Transfer points
● Haulages
TE ● Transformers
P ● Pumps
Sd.D ● Emergency sand dumps
St.D ● Emergency stone dust dumps
DWT ● Emergency drinking water tanks
● Face numbers shown brown when working

NOTE
Features marked ● must be shown on all plans. The inclusion of other features is optional but the above symbols must be used when any such features are shown

FIREFIGHTING INFORMATION

FS Fire Station
Shall contain as a minimum not less than 250m of hose with couplings two branch pipes, one dividing breach and one collecting breach, a supply of portable extinguishers, a supply of sand or stonedust and a supply of sandbags.

H Fire Hydrants
Shall be placed as follows:- (a) 20 - 30m approx. on intake side of all conveyor delivery points, main junctions, engine rooms, substations and fanhouses (b) at intervals not exceeding 230m along fire ranges (c) in close proximity to each coal face.

RV Water pressure reducing valves
V Water stop valves
—75 Pipe column with diameter (mm)
FP Flow pressure, shown at intye end of fire mains in 1000 N/m²
FFF Fire Fire-points
Shall be provided at fire hydrant position (c) above and shall contain sufficient length of hose to reach the ends of the coal face and one branch pipe.

FP Fire-points
Shall be provided at fire hydrant position (a) above and shall contain at least 5 lengths of hose and one branch pipe

E Portable extinguishers
Shall be provided on intake side of all fixed electrical apparatus and each unit shall consist of:
1 CO₂ extinguisher, 1 foam extinguisher, a bucket of sand or stonedust
or 2 dry powder extinguishers and a bucket of sand or stonedust.
Where considered necessary due to quantity of equipment, two or more units may be provided in which case the symbol E would be repeated for each unit (EEE)

SOURCE OF WATER SUPPLY
Hetty Shaft Surface pond gravity outlet 1 750 000 l. capacity.
Bertie Shaft Surface pond gravity outlet 200 000 l. capacity.

NATIONAL COAL BOARD SOUTH WALES AREA
TYMAWR LEWIS MERTHYR COLLIERY
YARD, UPPER NINE FEET AND FOUR FEET SEAMS
FIREFIGHTING AND RESCUE PLAN
SECTION

SCALE : 1/2500

Plan gridded in 100 m squares

Information supplied by Safety Engineer
Plan revised 26 April 1983

Boring the Rippings in 1966. Arch Girders (Rings) support rules: The rings must be set on a proper foundation and where the floor is soft must be set on suitable foot-blocks. Rings must be made tight to the roof and sides and fixed to their neighbours to ensure that none can become displaced. Struts set between the flanges of the rings serve this purpose except near the end of a line of ring, where some form of tie is necessary.

Sentry Appointment

INSTRUCTIONS

BOREHOLE ➡

A 'Trabant' Gel Stemming Ampoule, instructions:
1. With knife slit ampoule skin 2 or 3 times each side where marked.
2. Put blunt end of ampoule into borehole.
3. Push ampoule up to explosive charge.
4. Ram ampoule with stemming rod to force out gel to seal borehole.

In 1963 caravans were made available for fifty colliery workers at Llwyncelyn.

The Ripping Lip in 1966. In the photograph can be seen newly erected steel rings lagged with timber, hydraulic Duke props supporting the roof and ripping lip in a supply road. The tension box of the belt conveyor is also clearly in view. The supply road supplies the face and district with timber, props, rings, rails and sleepers. The road was ripped on the afternoon and night shifts.

A timber cog supporting and controlling the roof at the face in 1966. French timber and Duke props were used to support the roof. The coalface was hand cut with the aid of pneumatic picks and was advanced 4ft 6in on a forty-eight-hour cycle. The coal was filled on the morning shift only, with the cycle of operations arranged so that the face conveyors were moved forward on alternate afternoons. There was a similar alternation in the waste drawing and packing operations done on the afternoon shift.

Coal Cutting Lewis Merthyr 1906. This early type of disc coal cutter was driven by compressed air. When these coal cutters were introduced the proportion of large coal increased by about fifteen per cent. In this early period it was only the Hafod seam that was worked by coal cutters.

Name of Seam Depth	Great Western yards	Lewis Merthyr yards	Clydach Vale yards	Mardy No.1 Pit yards
Fforest Fach	33	-	-	-
Rhondda No.2	79	41	-	-
Rhondda No.3	148	115	61	-
Hafod	204	170	143	59
Abergorki	232	199	183	133

A housecoal curling box at Lewis Merthyr Colliery. A 'curling box' is a curved shaped metal box used to carry coal from the face to the Dram. The size of a housecoal curling box was usually approximately 17.5in long, 12in wide, 8in deep and held 20lb of coal. Will Salter had the sack (his employment terminated) in 1934 for filling dirty coal (coal and stone).

Tom Neate prop setting at the North 91 coalface in 1978. The Duke props shown at the coalface has a yield load of 21 tons. Following the publication, in 1973, of *PLAN FOR COAL*, a massive capital investment programme was launched by the NCB, aimed at securing the nations vital energy supplies well into the next century. For South Wales alone, over £100 million was earmarked for major modernisation projects; to streamline long-life pits; to locate and log new coal reserves; to create new mines and to link established neighbouring pits into single high-production units-Britain's guarantees in a fuel-hungry future! This colliery was one of those involved in the massive programme.

The Coal Cutter in the North 91 coalface in 1978. In 1978 the annual saleable output was 130,000 tons; average saleable weekly output 2,952 tons; average output per man/shift at the coal face 2 tons 16cwt; average output per man/shift – overall 1 ton; average weekly washery throughput 2,952 tons; total capital value of plant and machinery in use £183,000; coal winding capacity per cage wind 6 tons 18cwt.

Deputy Idris Reed testing for gas (Firedamp) in 416s face, 29 April 1983. With a Thomas & Williams Garforth lamp. When firedamp is mixed with air in certain proportions the mixture is explosive, which means that once it is set alight the flame will pass in a flash in all directions throughout the gas. Because of the dangers of an ignition of firedamp, Regulations require testing for it during a Deputy's inspections and before shotfiring and lay down rules for the provision of detectors. The limits of flammability of firedamp are: lower limit five per cent, upper limit fifteen per cent. There are many ways in which it is possible to set alight such a mixture of gases in a mine. Examples are: by flames – matches, cigarette lighters, damaged safety lamps, mine fires; by overheated surfaces – lamp gauzes, filaments in broken electric bulbs, brake drums, conveyor driving heads and tension ends, rollers with defective or unlubricated bearings. By friction sparks – sparks produced when rocks such as sandstone or pyrites are struck glancing blows by steel such as cutter picks or when light alloys are struck glancing blows by steel. By electric sparks – produced from switchgear, damaged cables, faulty electrical equipment; by static electric charges, which may build up in compressed air lines and discharge at nozzles or leaky joints. By explosives – heated particles projected from shot-holes and the ignition of firedamp in breaks or cracks in shot-holes. Firedamp may be ignited through: (a) Damaged safety lamps, (b) Defective or unlubricated bearings on conveyor idlers, (c) Friction sparks from cutter picks, (d) Damage caused to electric cables, (e) Heated particles projected during shotfiring.

In 1983 support to the roof was provided by Gullick Dobson chocks, which were set at regular intervals along the coalface. Each support was controlled by a number of valves, which admit hydraulic fluid through hoses into the various parts of the support. This allowed the support to be set to the roof, yield to the movement of the roof and floor, push forward the conveyor, be released and finally, move forward into its new position. These supports have been developed over many years from earlier types of support used on the coalface. Hydraulic power is strong and silent and accidents can occur if it is not used with caution. To ensure that powered supports operate correctly at all times, the basic rules must be observed. All powered supports must be set correctly and a defective support reported to the Deputy immediately.

The Pithead Baths on 10 May 1983. The photograph includes: Dickie Pember, and Jimmy Groves taking a shower in the colliery pithead baths. The official opening ceremony of the Lewis Merthyr Colliery pithead baths took on Saturday 4 October 1952 at midday. With R. Richards, ME, colliery manager (in the chair), J. Shenton, ME area general manager introduced Sir Hubert Houldsworth, QC, DSc, Chairman, NCB. Sir Huber Houldsworth and H. Lyn Jones, MC, Deputy Chairman, South Western Division made speeches on behalf of the NCB and W. Paynter, President (South Wales Area NUM), and I.R. Thomas, MP made speeches on behalf of the NUM. A vote of thanks was given to Sir Hubert Houldsworth, and visitors J. Jones, Miner's Agent (NCB), Cllr W. Cann, Chairman, and Lewis Merthyr Lodge of the NUM.

Colliery Canteen Staff in the 1960s. The meals in the canteen were always good value and a very high standard was achieved considering the surrounding environment the staff worked in. The staff served breakfast, dinners and freshly made sandwiches that the NCB issued when working overtime (doubler). The canteen also sold excellent quality towels and napkins at a reasonable price, single cigarettes, not forgetting the 'bob screws' of chewing tobacco to keep the dust at bay.

 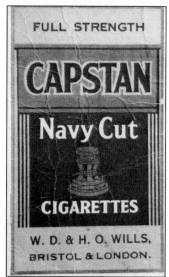

Left: Wild Woodbine Cigarettes. Available in the colliery canteen or from Mr Davies the sweet shop, at 4d for 10 or ½d each during the Second World War. His shop was opposite the main colliery gate and was open very early in the morning, working three shifts and eventually closing late at night. These very popular and inexpensive cigarettes were a favourite with smokers. *Right*: Capstan Full Strength Cigarettes. Available in the colliery canteen or from Mr Davies, at 6½d for 10 or ¾d each, where the miners, going to work, would buy one cigarette and light it from a small gas jet on the shop wall. Capstan cigarettes were strong and a little expensive but still a favourite choice with many smokers.

Left: A Colliery Newspaper Cartoon in 1947: 'How far to the Oval, mate?' The NCB coal magazines and the *Coal News* were popular with the miners and often available and read in the colliery canteen. In 1947 the *Coal* newspaper was 4d. *Right*: Primitive coal processing: hand sorting in the washery (a modern coal preparation plant usually found at most Pits). Sometimes called the Boris Karloff dungeon because of the diabolical working conditions.

Bathed and Going Home in 1958. The pithead baths were designed to give a first-class service to the workforce. There were separate rooms with lockers for clean clothes and pit clothes, ensuring that the former do not become soiled. The Baths adjoined the locker rooms, so that it was directly and easily accessible. Clothes hung in the lockers were dried by a current of warm air, which is continually passing through them. Facilities were available at the pit entrance of the baths to brush and grease the miners' boots and to fill their drinking water jacks. There was also an up-to-date Medical Centre in which serious and minor injuries could be dealt with promptly.

NATIONAL UNION OF
MINEWORKERS—(South Wales Area)

OFFICIAL

PICKET

G. REES, General Secretary.

National Union Of Mineworkers (South Wales Area) Official Picket Card. Authorised by George Rees, General Secretary. An experienced person, preferably a trade union official who represents those picketing, should always be in charge of the picket line. He should have a letter of authority from his union, which he can show to police officers or to people who want to cross the picket line. Even when he is not on the picket line himself he should be available to give the pickets advice if a problem arises. Throughout November 1971, a rash of unofficial strikes over pay disputes caused great unrest in the Welsh mining communities. This industrial action brought matters to a head and a strike was called on 9 January 1972. The national strike, the first since 1926, resulted in the whole of the South Wales Coalfield being brought to a standstill.

Lewis Merthyr Colliery in 1983. It was to be almost two months before coal was again raised, but the dispute, which had a devastating effect on British industry, saw the miners return to work as victors. To some, it was in some small way a vindication of their fathers' and grandfathers' suffering after such a humiliating defeat forty-six years earlier. The strike had shown that, despite the increased use of oil and nuclear power as alternative energy sources, the nation's prosperity still relied heavily on coal. A further strike in 1974 again saw the union locked in a dispute, which ultimately brought down the Heath Conservative Government. For the mining industry, the strike was a fight to save jobs. However, perhaps many more view the loss of jobs in the coal industry as a small price to pay for an end to the terrible toll of human life, the suffering and the desecration of a once beautiful landscape, which were hallmarks of an era when 'Coal was King'.

Preparing to fill and make safe the Bertie Shaft in 1987. On 21 February 1983 miners at the Lewis Merthyr Colliery refused to come up the pit after the day shift and began their stay down strike in protest at plans to close the colliery. The men came up the pit to a hero's welcome after a four and a half-day stay down strike, which prompted an all out strike by South Wales miners. Sadly the courageous fight to save the Pit was lost. Lewis Merthyr Colliery and Tŷ Mawr Colliery were both closed in 1983 by the NCB.

On 12 April 1918 Thomas Richards, General Secretary of the South Wales Miners' Federation, issued the following notice.

Military Service Comb-Out

South Wales Miners' Federation, 22 Saint Andrew's Crescent, Cardiff, 12 April 1918.

Acting upon the decision of the Conference held at the Cory Hall, Cardiff, on the 11th inst., the Council appointed a Committee of its members to act with the National Service representatives in arranging and supervising the methods to be adopted in the Comb-out of the quota of workmen from the South Wales Mines.

The Committee have now ascertained the correct number of men to be taken as the quota of the South Wales Coalfield for the 50,000 to be Combed-out of the mines of the kingdom, and have deducted from the total the number of Class A men who have been Combed-out or have voluntarily enlisted since January 1st, 1918. They have also allocated the number still to be obtained from each Colliery. The number to be taken from your Colliery can be ascertained from the miners' Agent of each District.

Some of the workmen having received calling-up notices, which expire on Saturday, the 13th inst., in some cases, and on, Monday, the 15th inst., in others, it has been arranged that the workmen in both cases shall not be expected to present themselves at the Military Stations until Tuesday, the 16th inst.

Committees are empowered to give full consideration to appeals from the men called up for medical examination and are authorised to recommend the cancellation of a calling-up paper on any of the following grounds:

(a) That the workman was under 18 years and 8 months of age on January 1st, 1918, or was over 25 years of age on that date.

(b) That the workman is the main support of the family.

(c) That the family has already contributed a reasonable proportion of its members for Military or Naval Service.

The Committee having decided for any of the foregoing reasons to recommend that a workman should be granted exemption, shall immediately forward his name and full particulars with calling-up papers and setting forth the grounds for their recommendations to the Miners' Agent for the district, who will take the necessary steps for presenting the case to the Military Authorities.

It must be borne in mind that every workman exempted through the recommendation of the Committee will place upon the Recruiting Authorities the necessity for calling upon another to complete the quota decided upon for the Colliery.

In cases where the employers make an application for the exemption of any workman on the ground of indispensability, the Workmen's Committee are entitled either to agree or dissent from such application. If they agree, they shall countersign the application made by the employer. If they disagree, their grounds of objection should be sent in writing to the Miners' Agent for the district.

(Signed) Thomas Richards General Secretary

Below is the price list and rules referring to a newly developed coalface at Lewis Merthyr Colliery in 1951.

Board of Conciliation for the Coal Trade of Monmouthshire, South Wales and Gloucestershire Pit Conciliation Scheme.

Reference to Umpires of Unsettled Question.

Lewis Merthyr Colliery Six-Feet Seam, Hand Cut and Conveyor Price List. Date of hearing 13 July 1951.

We have examined all parts of this Seam open for inspection and we hereby determine that the terms of the Price List shall be as follows:

1. Thickness of Seam Price Per Square Yard
    6ft   8s 6d (42½p)
2. The thickness of the seam to be taken by measuring from floor to roof. For every one inch decrease on the thickness of the seam 5d to be deducted and for every inch increase in thickness 5d to be added up to 6ft 6in. Above 6ft 6in thickness 75d per inch to be added.
    The above prices are for clean coal only and include the erection of all face supports, whether timber or steel, required on the face to comply with the Coal Mines (Support of Roof and Sides) Regulations, 1947. The prices cover the shifting and doubling of all props required in the face and the withdrawal of any props in the face side of the conveyor, handling and stowing clear of the conveyor all rubbish in the seam and made in the face, keeping clean conveyor belts, troughs and rollers, so as to be in a reasonable running condition.
3. Double Timber, notched and lagged, when ordered by the Management per pair 1s-6d.
4. Cutting rib per yard forward 1s-6d.
5. Starting and stopping Jigger conveyor engine by collier per day 1s-6d.
6. Doubling posts or steel props behind conveyor each 6d.
7. All coal to be placed on the conveyor and properly cleaned.
8. No coal to be thrown in the gob and it shall be the duty of each man to clean his stent properly before the shifting or turning of the conveyor and place all loose coal on it.
9. Suitable timber or steel props, as far as practicable, will be placed on the conveyor for the colliers' use.
10. Turning of the conveyor to be done by the Management.
11. All work done behind the conveyor to be paid for by arrangement between the Management and the workmen.
12. The men agree to work off all coal in a straight line so that there shall be no ribs on the face.
13. In the case of workmen being absent, the men in work agree to work off the absentees coal but the management have the right to make up the full complement of men.
14. It is the agreed duty of all colliers if they have cleared their own stents before finishing time, to assist as directed by Officials to clear off any bunches of coal left in the face.
15. Any abnormal conditions shall be met by mutual arrangements between the Management and workmen.

Date 1 October 1951
Signed   W.T. Woods
         Joshua Thomas
         Sam Garland

Lady Lewis Colliery in 1906. In 1891 the Lewis Merthyr Consolidated Collieries acquired the Llwyncelyn Colliery, Porth, in 1904 the company sunk the Lady Lewis Colliery a mile to the North East in the Rhondda Fach, in 1909 the company also sunk the Rhondda Main Colliery, Ogmore Vale, the colliery was plagued by serious water problems causing the colliery to close in 1924. The Meredith Level in the Rhymney Valley was acquired by the company and on 1 January 1947 the day of nationalisation, Cymmer Colliery, Porth also became part of the Lewis Merthyr Collieries, which was used for pumping mine water.

The scene following the explosion, 14 October 1913, at Senghenydd. In 1905 the Lewis Merthyr Consolidated Collieries Ltd acquired the Universal Colliery at Senghenydd, which was later to suffer the worst ever mining disaster in British history. The first of two dreadful explosions struck Senghenydd at 5.00 a.m. on 24 May 1901 when thirty-one men and boys were killed in the east side of the mine about 700yds from pit bottom. Although the ventilation continued to be very good, the heavy and extensive falls of roof greatly hampered rescue efforts. The second explosion occurred on 13 October 1913 when 439 men and boys were killed.

The Great Western Colliery, Hopkinstown in 1904. The colliery height was 293ft OD and was sunk in 1848 by John Calvert to the Rhondda No.3 seam at a depth of 149yds. The colliery was leased to the Great Western Railway Co. for ten years in 1854 and sold to the Great Western Colliery Co. in 1866. Three more shafts were sunk in the 1870s. On 12 August 1892 an explosion killed sixty-three (from the actual report) men and boys. The three winding shafts were used by the mines rescue team and the rescuers.

The Great Western Colliery in 1904. On 4 December 1874 the accident reports show that thirty-five-year-old stoker H. Marslyn was killed by falling down the pit while engaged in disconnecting drams. On 5 July 1875 the accident reports show that thirty-four-year-old fireman (colliery official) J. Prosser was killed by a fall of stone. On 1 January 1884 the accident reports also show that twenty-two-year-old mason, Griffiths Davies and twenty-three-year-old mason, Alfred Cornish, were killed by a fall of stone. The Great Western Colliery closed in 1923.

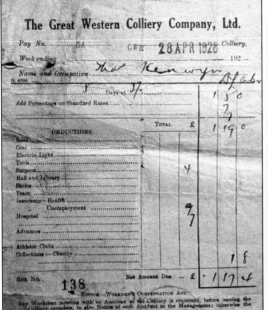
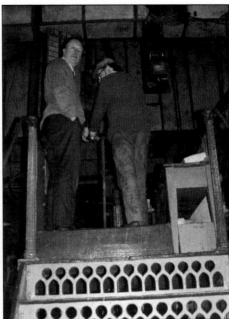

Left: Great Western Colliery Pay Docket, 28 April 1928. The take home pay was £1 17s 4d for five days work. *Right*: The Hetty Colliery Winding Engine House, Hopkinstown on 16 April 2000 with Martin Doe (Friends Secretary) and Brian Davies (Pontypridd Historical Centre Curator). Barker & Cope of Kidsgrove Staffordshire built the steam-winding engine in 1875 and was originally built with two 40in cylinders with Cornish valves, it was later fitted with new 36in cylinders and piston valves. The drum has a diameter of 16ft and originally held a flat rope.

The Hetty Pit in 2000. The Hetty shaft was sunk by the Great Western Colliery Co. in 1875 to the Six-Feet seam at a depth of 392yds. The shaft was 16ft in diameter and 392yds deep. Brian Davies, curator of the Pontypridd Historical Centre, has restored the winding engine, which is now in working order and will be open to the public in the near future.

Tŷ Mawr Colliery, Hopkinstown, in 1983. The colliery was sunk in 1923 by the Great Western Colliery Co. The average seam sections were; Two-Feet-Nine: 58in, at a depth of 349 yards, Four-Feet: 77in, at a depth of 365yds, Six-Feet: 81in, at a depth of 392yds, Red-Coal: 40in, at a depth of 409yds, Nine-Feet: 68in, at a depth of 421yds, Lower-Four-Feet: 76in, at a depth of 442yds, and the Five-Feet at 68in, at a depth of 475yds.

Tŷ Mawr Colliery Screens in 1983. In 1954 with a manpower of 860 the colliery produced an annual output of 250,000 tons, in 1955 with a manpower of 850 it produced 230,517 tons, in 1956 with a manpower of 873 it produced 221,882 tons, and in 1957 with a manpower of 900 it produced 240,297 tons.

Tŷ Mawr ripping lip in 1983. The ripping lip is the removal of stone from above the seam to create a higher heading at the entry to the coalface. In the photograph the ripper can be seen standing on a temporary staging ripping down the stone ripping lip using a pneumatic pick (one of the many causes of vibration white finger) in preparation to erect steel rings which are then lagged with timber. The road was ripped on the afternoon and night shifts.

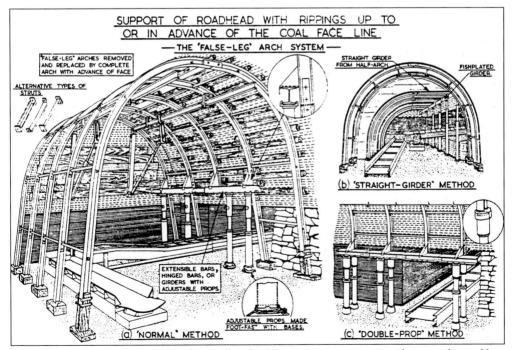

SUPPORT OF ROADHEAD WITH RIPPINGS UP TO OR IN ADVANCE OF THE COAL FACE LINE
— THE 'FALSE-LEG' ARCH SYSTEM —

The diagram shows the support of a roadhead with rippings up to or in advance of a coalface line. The false-leg arch system. A conveyor in the gate road carries the coal outbye. The supply road supplies the face and district with timber, props, rings, rails and sleepers.

Tŷ Mawr Colliery Pitmen in 1983. The photograph includes: Dick Pember and Glyn Allsop (Engineer). In 1958, the NCB invested £1.2 million in a surface and underground reorganisation, to merge and streamline Lewis Merthyr and Tŷ Mawr. The two mines employed 730 men and produced an average annual output of around 132,000 tons. The working area was almost twelve square miles and incorporated around twenty-one miles of underground roadways, in which there were over three and a half miles of high-speed belt conveyors in daily use.

The Last Dram 21 June 1983. The photograph includes: Alan -?-; A. Morris; Collin -?- (overman); Dai Rees (hitcher); Dai 'The Lampman'. The colliery was owned by the Powell Duffryn Steam Coal Co. prior to nationalisation in 1947. The last dram of coal was raised on 21 June 1983. Tŷ Mawr Colliery was merged with Lewis Merthyr Colliery in 1958 and was closed in 1983 by the NCB. There was always tremendous courage and staunch camaraderie with the South Wales Miners in the deep, fiery and dangerous pits.

There were no greater mineral treasures than the riches of coal found in such abundance in the valleys of the South Wales Coalfield.

Two

Trehafod Mining Village in the South Wales Coalfield

One of the twelve counties of Wales, the old county of Glamorgan consisted of ten ancient Hundreds – Cowbridge, Dinas Powys, Kibbor, Llangyfelach, Miskin, Neath, Newcastle, Ogmore, Senghenydd and Swansea.

Each Hundred contained parishes, 126 in total. The Hundred of Miskin contained seven of these parishes: Aberdare, Llantrisant, Llantwit Fadre, Llanwonno, Pentyrch, Radyr and Ystradyfodwg.

Of these the parishes of Llanwonno and Llantrisant play an important role in the history of Trehafod. Before the advent of coalmining there was no Trehafod, or at least the locality now known as Trehafod, was Hafod-Ucha, Canol, Fawr and Hafod. All of these were in the parish of Llanwonno. Most of the land was owned by the William family of the Glog (Llanwonno), with the deeds dating back to 1614.

The other areas were Eirw, Eirw Isaf and Eirw Uchaf, all on the south side of the river, and in the even more ancient parish of Llantrisant. Both the Eirw and Hafod areas were sparsely populated and primarily farmland. As the area became more populated it became Hafod, and in the latter part of the 1800s Trehafod.

However the old boundary lines still existed, and, even today, there is different political representation for the local Council, Welsh Assembly and Parliament, for different parts of Trehafod. Situated at the lower end of the Rhondda Valley, surrounded by mountains-which have seventeen streams running down into the River Rhondda, Trehafod is approximately two and a half miles from the town of Pontypridd and fourteen miles NNW of Cardiff. The English translation of Trehafod is: *Tre*(f) meaning town and *Hafod* meaning Summer Dwelling.

On the local farmland, one or two farmers discovered that there was coal on their land. They worked the coal in shallow primitive levels under their homes for several centuries. However, in 1730 the first employer pioneer was Dr Richard Griffiths of Llanwono who opened a level at Gyfeillon (a stones throw from Trehafod). In 1809 Jeremiah Homphrey leased some land in Hafod for coalmining. The land was on the north side of the river and was later developed into the Hafod Level. Coal that was mined needed to be distributed, and a tramroad was built from Hafod to the Glamorganshire canal at Treforest (the canal had been opened in 1794, running from Abercynon to Cardiff).

In 1835, land owned by William Crawshay was turned into a level called the Gwaered-Yr-Eirw Level, on the south side of the river, alongside the tramway, where the present path leads down to Barry sidings.

This level was eventually closed in about 1914. Following this level many others were opened, one opposite the present site of the Heritage Park Hotel, one on the hill leading up to Woodfield Terrace, one on Parish Hill, three others on the old tramway and in 1844, Lewis Edwards developed the Nyth Bran Level.

The Coed Cae Colliery was opened in 1850 by Edward Mills on the present Heritage Park Hotel site. Whilst the Llwyncelyn Colliery was near the Britannia Hotel. This colliery was owned by D.W. James but in 1880 he sold the venture to Edward Thomas, who also owned the Coke works. The coke works was situated between the two collieries. The Hafod site was now under the ownership of Thomas Jones, who also owned the Ynyshir Colliery.

In the early nineteenth century most of these Welsh coalmines had been small in scale. The majority of coalowners had been Welshmen, often nonconformist who lived locally and enjoyed a close relationship with their employees, often sharing the same liberal thoughts and attending the same chapels.

However by the 1880s, the growth in the coal industry had elevated the coalowners to new heights and most of the original owners had sold their interests. In 1883 the Coalowners Association of South Wales had been formed, promoting a new capitalist class. This association included Powell Duffryn, Cambrian, Ocean Group and the Lewis Merthyr Combine.

By 1880 the Bertie shaft had been sunk and in 1890 the Trefor shaft. Sir William Thomas Lewis, later Lord Merthyr of Senghenydd owned the Lewis Merthyr Group. As well as the collieries he also owned much of Cardiff Docks and he lived in baronial splendour. His attitude towards the coal industry was almost feudal, as he ruled the coalowners association in an autocratic fashion.

He chaired a committee of five coalowners and five miners representatives led by William 'Mabon' Abraham, which negotiated in 1875 a sliding scale for wages. Under this scheme a change in price of 1 shilling a ton would involve a change of 7.5 per cent in miners wages. The scheme was controlled by the owners and miners had to rely upon the word of the owners as to how much a ton of coal fetched. The scheme was fraught with difficulty and it led to a stoppage of work in 1898.

Following this miners were locked out of work for six months, poverty was the order of the day; soup kitchens were in existence. In September of that year miners returned to work in what can only be described as a complete victory for the coalowners. Other strikes occurred particularly the strikes of 1906 and 1908, but the pursuit of coal and profits was relentless, as was the growth of the Rhondda Valleys including Trehafod.

By 1913 coal production in South Wales had reached its zenith – 56 million tons of coal, with 30 million tons being exported. However productivity had already began to fall from 309 tons per man-year in 1893 down to 222 tons per man-year in 1913. This was partly due to geological difficulties.

Throughout the periods companies became bigger and bigger in 1929 the Lewis Merthyr was now part of Welsh Associated Collieries, and in 1935 it became part of the largest coal group in the UK, the Powell Duffryn Group known to many as the 'Poverty and Dole' Group. The local collieries remained with this group until the famous day of 1 January 1947 when all pits passed into public ownership. The local collieries forming part of the NCB's South West Division.

In 1841, recognising the needs for the transportation of coal, the Taff Vale Railway Co. extended their steam train line from Pontypridd to Dinas, and eventually throughout the Rhondda. Each colliery had its own sidings; consequently coal was more easily transportable. Much of the coal was taken to Cardiff Docks and the first cargo of Rhondda coal was shipped to France in 1842.

The Hafod collieries were well served by this line. As well as coal, passengers were also able to use this important means of travel. Hafod had a railway station, situated on the south side of the river and on the present Rhondda Heritage Park site. One of the first stationmasters was Thomas Meredith. However, as the collieries developed, and they needed additional space, the station was resited at its present location in Trehafod Road.

In 1884 the Barry Docks Bill was passed through Parliament and by 1889 coal was being transported directly to Barry. This was made possible by the construction of a new branch line, and by the construction of a viaduct, bridging both the road and the river at the eastern end of Trehafod. New sidings developed and were known as Barry Sidings. The sidings had Engine sheds, and there were many hundreds of wagons filled with coal awaiting transportation. The line progressed through a new tunnel in the Maes-Y-Coed area of Pontypridd.

In 1890 approximately 2 million tons of coal were carried into Barry. By 1900 this had increased to 2.5 million tons of coal in addition to many thousands of tons of coke. Passenger trains also used this line – carrying many hundreds of families on a day out in Barry. The first Monday of each month was a miner's holiday, and the trains bore the letters 'M.D.', Mabons Day, a day, which had been negotiated for by the leader of the men. This holiday was taken from the miners following the 1898 coal stoppage, which lasted for six months.

1922 saw the takeover of the rail network by the Great Western Co. Over the years, as exports of coal diminished, the sidings were used less and less and the line was eventually closed. The viaduct was dismantled in 1956, but some of the support pillars can still be seen.

In 1851 the population of Trehafod was 370 as shown below:

Age	Hafod		Eirw		Trehafod		Total	
Years	M	F	M	F	M	F	M	F
0-5	1	7	1	1	43	38	45	46
6-10	5	1	0	0	27	14	32	15
11-20	2	2	1	1	32	21	35	24
21-30	3	6	1	1	24	24	28	31
31-40	2	0	1	0	32	21	35	21
41-50	4	2	1	0	21	11	26	13
51-60	2	2	2	1	0	1	4	4
61-70	0	0	0	0	4	3	4	3
70+	0	2	0	0	2	0	2	2
Totals	41		11		318		370	

It is interesting to note that only four people were aged seventy or more, and that the oldest inhabitant was seventy-four years of age; indeed, only nineteen people were aged fifty years plus. These facts are hardly surprising as most of them had moved into the area seeking employment. They came from many parts of the UK, including Middlesex and Kent, but the largest contingents were from the rural areas of Monmouthshire, Carmarthenshire, Cornwall, Breconshire, and Somerset.

According to the 1851 census, the 370 people were housed in approximately fifty-six houses, two public houses and a few farms. On average there were six people to a house (currently there are about 320 houses).

Because of a shortage of suitable housing for single people, many households were taking in lodgers. Although making sleeping arrangements difficult, the rent paid by the lodgers certainly helped the family finances. Many of the new houses were, typically, two/three bedrooms, an outdoor toilet and certainly no running hot water or central heating.

Mervyn John

A painting of Berw-Y-Rhondda by Henry Gastineau in 1805. Berw-Y-Rhondda, near Trehafod, was situated at the gateway to the Rhondda Valleys. Before the influx of the industrial revolution the Rhondda Valleys and surrounding areas enjoyed a pastoral tranquility, with bubbling fresh water streams and heavy woodland. The village of Trehafod grew on the estate of the Morgan family in the area known as Hafod. Hafod is the name given to the area of land stretching from the Great Western Colliery in Hopkinstown to Porth, ancient entrance to 'The Holy City' of Dinas. (K. John, 1993)

Eirw Farm in the early years of the twentieth century. The Hafod community would have consisted of the Morgan family, owner of the large Hafod Fawr estate, a number of smaller tenant farmers, craftsmen, a small group of labourers, servants, apprentices and shepherds. The community would have been self-sufficient, with their existence dependent on the traditional farming of sheep and some cattle.

The Old Sheave above Trehafod was used to transport stone to build the houses in the locality. Along with Hafod Fawr there existed a number of smaller farm farmsteads, such as Hafod Fach, Nyth Bran, Hafod Ganol and Daren-Y-Pistyll. Along the lane where Hafod Ganol and Hafod Uchaf stand is the route taken by the monks and their sheep travelling from Neath or Margam Abbey to the monasteries at Ynysybwl. The Drovers Roads were also early routes for transporting cattle from village to village. The road ran from Brecon to Llanwynno, passed through Trehafod and on to Llantrisant and Cardiff.

Hafod Fawr (Dr Morgan's), is the mansion at the east side of the river Rhondda and was home of the Morgan family. The whole of the Morgan estates were a game reserve and there were often as many as twenty guns on a shoot. These would have been attended by lords and nobility from all over the country. A dozen or so men were employed in the mansion grounds with twenty or thirty horses kept for use by the family and their many guests. It is documented that Miners Leader Mabon was a friend of the Morgans and a frequent visitor to Hafod Fawr. The Mansion's tower overlooks the Lewis Merthyr Colliery site and lands where several thousand workers would have been employed.

Many of the houses were built from stone from the local quarries. Trehafod had, at that time, four working quarries owned by William Phillips. One quarry was situated slightly to the west of Rheolau Terrace, one at Daren-Y-Pistell, one to the east of Thomas Place and one on the old tramway. They were in the main 'closed' by 1910, but a new one had opened in Cwm George. Most house building had been completed by 1919 and only a handful of newer properties have been built since that time. Most of the properties have now been fully modernised. (K. John)

Siloam Welsh Calvinistic Methodist Chapel during the flood in 1938. The photograph includes Mr Rees, the caretaker. Siloam Chapel was built in 1849. With the growth of the new community came the growth of religion in Trehafod. The number of places of worship indicates the religious intensity of the village people. At chapels, Sunday schools were well attended as they acted as a form of education for the villagers' children. The first school in Trehafod was a private school situated at the top of Sant's Hill with today's Hafod Primary School being built in 1878. The chapels also ran annual trips to the coast. Only those who were regular attendees were able to go on these trips thus enticing the children into chapels. Chapels such as Bethel, Bethesda, Penuel, Siloam, Zion and Wesleyan were non-conformist, although there was one Church in Wales. St Barnabas church was built in 1892 and is the oldest unconsecrated church in the Rhondda. There was also an unusual Pentecostal meeting place built from an old shop.

The schools also met in a chapel in Porth for the Eisteddfodau. The Urdd Eisteddfod was held at Ynysangharad Park at Pontypridd. This was the central area where all the major attractions took place, such as galas, fairs, processions and horse shows, and on some occasions more exotic animals such as elephants appeared.

The graveyard belonged to the chapel, which once stood on the ground next to it. Most of the graves date from the 1880s and show that few adults lived to over fifty. There are also many children's graves, which show the appalling infant death rate. The grave of the children of Benjamin and Elizabeth Jones, one of the few inscribed in English, is a good example. Six children died in the period 1880-1886. Only two of these reached the age of one. Working back from these dates shows that all the children were born in the autumn and by the following January Mrs Jones was pregnant again. A gap in the sequence suggests that there may have been a child born in 1882 that survived. Infant death was not confined to mining families. For the period 1892-1896, infant mortality in the Rhondda was 210 per 1,000 births; the average for England and Wales was 150 per 1,000. Though improvements were made in sanitation in this century, the poverty caused by the collapse of the coal industry after 1920 meant that adult and child death rates remained high until after 1939 (in 1937, the general death rate in the Rhondda was 13.4 per 1,000, whereas the average for England and Wales was 9.4 and for the English industrial towns 10-11 per 1,000). The first burial at Siloam was Dafydd Daniel, who died on 26 August 1850 aged twenty-one and the last internment was Thomas Jones who died on 17 July 1913, aged seventy-four. (Mervyn John)

70

The photograph includes Bethel Welsh Independent Chapel, Trehafod, built in 1863. In the years of the nineteenth century many members of Cymmer Church, those whose homes lay in the country between Cymmer and Pontypridd, gathered together for some years to worship at a small schoolhouse near Waun-Yr-Eirw. Their chapel was built on a site belonging to the Hafod estate in Llanwonno parish, leased for ninety-nine years for the total of ten guineas. When completed the building measured 36ft by 32ft inside, and cost £565, the work having been made possible due to the inexhaustible energy of sixty-year-old William Evans, who secured the site, instrumentally procured the raw materials, obtained loans for the projects and even laid much of the foundation himself completely free of charge and resting only when the building was completed.

The new chapel was opened on 27 and 28 September 1863, by Cymmer's minister, Revd Henry Puntan, who formed the church on 9 October with a total membership of twenty-nine. The brethren at Cymmer were overjoyed at being able to help found Bethel church, but rancour still lingered among those who resented the fall in membership, culminating in Revd Puntan being compelled to withdraw from charge of the church a few months later.

Consequently Bethel's first resident minister, Revd John Griffiths of Glantaff, was inducted on 11 and 12 May 1864, to share his pastorate with Ebenezer Glantaff, a situation that lasted until his departure in February 1869. That September, Methuselah Jones of Pentyrch, then studying at Milford Haven preparatory school, accepted a call to Bethel and was ordained on 6 December.

Presently the first minister was to transfer his services to Tynewydd and the second minister was to arrive from Bala College to replace him. Mr John Williams was ordained on 4 October 1875. In view of these upcoming services some preparatory interior renovations had been possible. The £150 costs, which they incurred, added to the £50 debt remaining from the building costs, an onerous burden for a small congregation.

The matter was dealt with effectively however. At the yearly meeting of September 1887, it was reported that only £5 remained outstanding. One final collection among the congregation that afternoon and the church was free from debt. But the members were not able to rest for all that. Industry was expanding rapidly around them the opening of Coed Cae, Hafod and Great Western Collieries, doubling the working population and with it Bethels congregation and Sunday school class. Investment in a new chapel became unavoidable when the search for a new site yielded a suitable plot of land, although rather steeply priced at 3d a square yard – £9 per annum. They opened the second chapel of Bethel here on 3, 4, 5 November 1883 at a cost of £2,200.

The first chapel, which stood nearby, then found use as a vestry until sustaining severe damage in a storm some years later; it has since been sold. The chapel became stronger than at any time previously, its work in all areas faring successfully. One member achieved a place in the ministry – Morgan V. Davies, who was ordained at Abergele in 1889, following preparation for the ministry at Pontypridd and Cardiff University. The chapel no longer exists.

Bryneirw Church in 1905. St Barnabas, opened in 1892. In 1887, a Sunday school was held in the surgery of Dr Naunton Davies, but was soon closed due to overcrowding. The Lewis Merthyr came to the rescue and placed the colliery reading room at the disposal of the church until 1892. £900 was raised to build a hall seating 250 people. This was built on the present site and was known as the Bryn Eirw Mission. The church continued to flourish and in 1930, the name was changed to St Barnabas.

St Barnabas Dramatic Society 28 June 1937. *Back row, left to right*: Miss G. Lewis, Mrs J. Morris, Miss I. Williams, Mr J. Smith, Mr T.H. Lloyd, Mr W. Lewis, Miss M. Davies, Mr J. Pugsley, Miss E. Davies. *Front row*: Mrs E. Hunt, Mrs J. Weaver, Revd D.I. Evans BA (Producer), Miss G. Smith, Miss O. Vaughan.

Bethesda Welsh Baptist Chapel in 2001. Bethesda Chapel was built in 1891. The background of the Baptist cause in Trehafod is to be found with Capel Rhondda, Hopkinstown. Capel Rhondda had run a Sunday school since 1853, but in 1882 an extra branch of it was opened, in a private house in River Street, Trehafod. In 1833, however, a small schoolroom was built (this became the chapel vestry) at a cost of £400, and the church of Bethesda formally established. Most of the members were originally from Capel Rhondda.

In 1885, the church accepted its first minister, Revd W. Bassett; at that time still a student at the Baptist College at Pontypool. Services were held in the vestry from some years, until the recent chapel was built in 1891.

Details of such of Bethesda's history are lost, as the church records were destroyed in the flood, which swept Trehafod. The following few facts are available, however during the First World War the church received the services of Revd John Frimstone, from North Wales, and he was followed at some point by Revd Davies and later by Revd Emlyn Lewis, whose ministry was shared with Tabernacle, Pontypridd.

Maximum membership was around 400, probably at the time of the Revival or shortly thereafter. In its heyday the church was very active locally in the music field, for instance, Mr Isaac Jones, led the Band of Hope and conducted the choir in operattas and local Gymanfas. During the strikes in the 1920s a soup kitchen was set up for the starving miners, which was run from the vestry, with the help of most of the members.

Since the Depression, however, the Church had suffered a very considerable decline. There had been no minister for several years, the congregation relying on the services of lay preachers, with communion from an ordained minister one afternoon a month. As membership had dropped to single figures, the church reverted to using the original vestry for its services, with the result that the main chapel was in a poor state of repair.

The similarity in frontage between this and the winding houses at the colliery is not a coincidence! Chapels were not meant to look like churches (to underline the idea that the established Church was 'fallen') and were built to secular designs, often by the same architects as those who designed the colliery. The chapel was the focus of life outside the colliery. Until the 1880s they were the only source of education for many families. From the chapels originated the choral and brass bands traditions of the Rhondda. They also held literary and poetry competitions (*eisteddfodau*). On the other hand, activities of which they disapproved, such as racing and street theatre, were stifled. The chapel is a characteristic yet vanishing feature of the Valleys. Doctrinal splits led to ever more being built, so that there was over-capacity at the best of times (Trehafod had six chapels at one time). Later the attractions of socialism and secular recreation began to affect chapel attendance. The chapel is now residential flats. (Mervyn John)

Zion English Presbyterian Chapel, built in 1900. The annual chapel outing to happy, noisy, bustling Barry Island or Porthcawl was eagerly awaited and talked about for several months before departure and the chapel's congregation would suddenly increase in numbers. Egg sandwiches, pop, crisps, 'Welsh cakes' and biscuits were prepared and packed ready to be eaten on the beach, weather permitting. A ride on the fair was kept to finish off the day in exhilarating fun. We would head for the dodgems and any other ride we could get our hands on, spending the last of our money we had left in our pockets in the penny arcade and purchasing a stick of Barry Island rock to eat on the way home. We forgot about the cost; it was only once a year.

Zion English Presbyterian Chapel built in 1900. The photograph, taken in 1959, includes, *left to right*: Mr Pugh, Howard McCabe, Neil Oliver, Haydn Pugh, Cynon Collyer, Valerie Bowler, Valerie Cooper, Annette Bowler, Miss G. Williams, Elaine Canard, Eileen Oliver, Cheryl Cooper, Jennifer McCabe, Pamela Davies, Cynthia Pugh, Pat Collyer, Kathleen Thomas, Mrs C. Hughes, Mrs McCabe, Mrs Canard, Mrs Pugh, Mrs Gaye, Mr Oliver, Mr Hughes, Mr Jarman, Revd Howells (new minister), Mrs Hughes, Mrs Llewellyn, Mrs Cooper, Mrs Amy Jones, Mr Crockett, Mrs Bowler, Mrs Keogh, Mrs Crockett, Mrs Rogers, Mrs Taylor, Revd Rogers, Mrs Davies, Mrs Glenys Pugh, Mrs Thomas, Byron Pugh.

Penuel English Baptist Chapel in 2001. Penuel Chapel was built in 1900. The chapels were full to capacity, but in the years of the depression in the 1920s attendances declined by at least sixty per cent. As more and more non-Welsh speakers migrated into Trehafod the Welsh chapels saw a further decline in their numbers. The ministers residing at Trehafod in 1910 were Revd Walter Daniel, Welsh Calvinistic Methodist; Revd John Frimstone, Welsh Baptist; Revd Benjamin Jones, English Calvinistic Methodist; Revd Idris Thomas Price, Bryn Eirw Church in Wales, and Revd John Williams, Welsh Independent.

English Wesleyan Chapel in 2001. The Wesleyan Chapel was built in 1904. The chapel was later to become a picture house, known locally as the 'Bomb and Dagger'. During these declining years there were tales of ministers having no stipend for months on end, and the chapels had no money for their upkeep. Other factors led to the continued demise of the chapels including the opening of cinemas in the 1950s, the Sunday opening of public houses in the 1960s, the advent of television and the euphoria of bingo halls. Unfortunately all of the chapels have now been closed, some of the buildings have been demolished whilst others have been converted for other uses.

Trehafod Social Club in the new millennium. The club was built in 1981 and was formerly the Royal British Legion Club. On 9 September 1999 the Royal British Legion Club celebrated sixty years; it had been formerly situated next to the Lower Eirw Bridge and Temple Buildings. The Royal British Legion Women's Section celebrated its 65th anniversary on Wednesday 10 October 2001. The clubs, pubs, and particularly the miner's workmen's halls and institutes, were the secular successor to the chapel in terms of social and educational activities when socialism replaced religion as the driving force in life. Earlier in the last century passions ran high between militant miners, who believed that a strike was the only effective weapon against the bosses, and moderates who believed that more could be won by negotiation.

Trehafod Bertie in 2001, formerly the Village Inn and the Non-Political Club (The Cock and Chick). The Bertie is now owned by Alan and Margaret Hamer. It was fully refurbished in 1999 and renamed the Bertie after the Pithead Winding House, situated in the Rhondda Heritage Park just 100 yards away. The Guest House offers an ideal choice for a drink, meal short or long stay in the heart of the Welsh hills with evening entertainment most weekends.

Trehafod Memorial Hall and Institute in 2001. Trehafod Memorial Hall and Institute was built in 1933. Four cottages in Trehafod Road were donated by Dr Morgan of Hafod Fawr. The two middle ones were demolished to build a Memorial Hall to accommodate a social club for miners and their families consisting of a cinema, a two-table snooker room, cards/bagatelle room and a library with the daily papers. The caretaker, Jacob Maddocks, lived next door. The cinema contains two war memorials, one for each World War, where a service is held every Remembrance Day. In 1963 a donation of £9,000 was given by the Miners Welfare Fund to form a Memorial Hall and Social Club. The cinema/concert hall is run by the Charity Committee who rent the hall to local groups, jazz bands, play groups, cubs and private functions. There is a business aspect of the hall run by the Social Club Committee and consists of a Bar and Lounge, pool/snooker and games rooms on the ground floor and on the first floor there is a Bar in the cinema/concert hall. Its full title is now Trehafod Memorial Hall and Social Club Ltd.

The Wayne Morgan Conservative Club Members in the 1950s. The photograph includes: Eddie Owen, Sam Bailey, Glyn Davies, Mr Thomas, Ray James, Trevor Owen, Glan Morgan, Charley Tuvey, Dai Goode, Paul Robathom, Jimmy Poole, Mr Lloyd, Mog Walters, Ernie Barrie, Idris Davies, 'Bomber' Clift, Phil Northall, Bill Northall, Mr Owen. Wayne Morgan, a character of the village, was one of the founder members of the chain of Conservative Clubs. There was one in Trehafod and from there he built up the clubs throughout the valleys. Morgan was instrumental in the Chartist and Rebeccarite movement, and from his foundry at the bottom of Trehafod the turnpikes were made. The Wayne Morgan is a Men Only Club. (K. John)

Lewis Merthyr Workmen's Hall and Institute Officials and Committee in 1955. *Back row, left to right*: Cllr W.J. Cann (Trustee), J.P. Jenkins, E. Beynon, B. Lewis, C. Poulsom, B.P. Bessant, R. Jeynes, C.R. Owen (Vice-Chairman). *Front row*: William Parry (Trustee), Henry Tyler (Secretary and Trustee), Richard Dentus (Chairman and Trustee), Bryn Rees (Trustee).

Lewis Merthyr Workmen's Hall and Institute, Billiard and Snooker Team, winners of Pontypridd and District Handicap Shield 1956/57 season. *Back row, left to right*: H. Tylor (Secretary), W. Lewis, I. Jones, G. James, G. Jones, D. Gower (Marker). *Front row*: J. Hill, G. Bishop, T. Crease (Captain), E. Hughes.

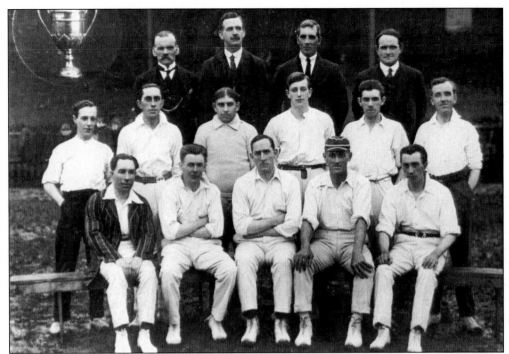

Lewis Merthyr Workmen's Hall C.C., winners of the 'Sugg Challenge Cup', 1922 season. *Back row, left to right*: J. Gibbon, T. Davies, W. Webb (Honorary Secretary), J. Roberts. *Middle row*: E.J. Morgan, Trevor Williams, Geo Hemmings, T.J. Hicks, W.A. Derham, Tom Vaughan. *Front row*: Jack Morgan, F. Cridland, G.F. Charles (Captain), W.A. Williams, Tudor Williams.

Lewis Merthyr Workmen's Hall Tug of War Team in 1938. *Left to right, back row*: I. Hooper, G. Healy, H. Withers, D. Davies, C. Grenfield, I. Owens, A. Davies, E. Jones, P. Chapman. *Middle row*: I. Stallard, H Churchill, E. Morgan, A Chapman, B. Bentrus, E. Lewis. *Front row*: E. Williams, I. Iles.

The 'Tump' Sports Team in the early 1900s. A Trehafod football team can be traced to before 1900. In the 1920s and 1930s there were teams in the village with names such as, Lewis Merthyr, Lewis Merthyr Celtic, Trehafod Ex-School Boys and one team made up entirely from one family named Downes. The old Coed Cae football field where the team played possessed very little grass and was made up of clay, stones and rocks with an uneven slope running down towards Pontypridd.

Lewis Merthyr AFC League Champions 1956/57 season. *Left to right, back row:* Ivor Simmons, Phil Jones, Johnny Martin, Cyril Lacey, Selwyn Evans, Trevor Moulding, Islwyn 'Bookie' Sheppard, Terry Wilcox, Brian Hole, Roy Hamer, Ken Richards, Stan Bull. *Middle row:* Cliff Bull (Secretary), Eddie 'Bunter' Hughes, Mervyn 'Toughie' Lloyd, George Davies, Dickie Hopkins, Phil Gimlet, Dai Jones. *Front row:* Tommy Pooley, John Davies. Professional footballers in the village included: E. Cornish (Bolton Wanderers and captain for Wales five times), Wallace Watkins (Bolton Wanderers), Tommy Hicks (Nottingham Forest), Tommy Clarke (Nottingham Forest), John Beynon (Aberdeen), Dai Daniels (Torquay), Tommy Jones (Bradford), Harold Rasmond (Crew & Ipswich), Tommy Evans, Henry Picton (Bournemouth), and School Boy Welsh International, Daniel Rees.

Lewis Merthyr Band at the South Wales Miners Gala, Queen Street, Cardiff in 1954. A.J. Cook, South Wales Miners Leader, is pictured on the Lewis Merthyr National Union of Mineworkers Colliery Banner. If you know the whereabouts of this Banner please telephone the Heritage Park.

Now in its 119th year, Lewis Merthyr is the oldest Brass Band in the Rhondda Valleys. Originally known as the Cymmer (Porth) Colliery Band, Lewis Merthyr adopted its present title in 1949.

However, it has been established that in the early part of the century, a band, originally called the Trehafod Recobite Band, bore the name of 'Lewis Merthyr' as it received support off the then prosperous, Lewis Merthyr Colliery. In the early 1900s, it changed from an all-brass band to a military type band. Press cuttings from that period showed a regular Sunday evening attendance of thousands of listeners to concerts held at Porth Park.

Lewis Merthyr has gained National honours by winning the Third Section and Second Section finals at the National Brass Band Championships, at the then Belle Vue, Manchester, in 1950 and 1951 consecutively, therefore gaining 'Championship' status in just three years. The band has represented Wales eight times at the National Brass Band Championships in the Royal Albert Hall. In 1985, having become Champion Band of Wales, the band represented the principality at the European brass band Championships held at the Tivolli Gardens, Copenhagen, Denmark.

For the past five years Laurence W. Harries BP Phil, MA (Open) has been the band's musical conductor. Laurence started playing a brass instrument at the age of six, with the then Glen Rhondda Band in Treherbert and at fifteen joined the Army as a musician. He studied at the Royal Military School of Music, Knellar Hall, as a pupil in Euphonium and later as a Student Bandmaster. After leaving the service, he continued his association with local bands in the Rhondda; ill health precluded him from playing a number of years ago, so he now concentrates exclusively on his conducting career.

The current longest serving member of the band is sixty-seven-year-old Gilbert Cook, who has been with the band for fifty years. Gilbert is musical conductor of the Junior Band.

The band is available for fetes, carnivals, charity concerts and fund-raising concerts. Any person interested in playing for the Band can contact any band member.

Rhondda has long been a centre of cultural life, the home of writers, actors, musicians mighty male voice choirs, brass bands and the Rhondda Symphony Orchestra.

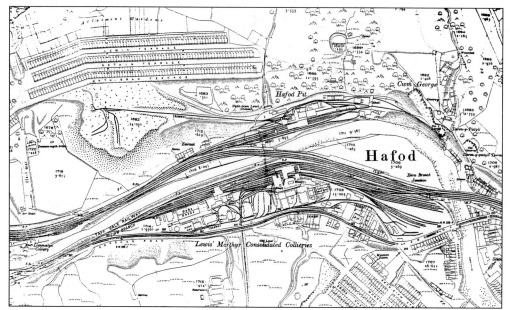

A Plan of the Lewis Merthyr Consolidated Collieries and Trehafod in 1919. Place names include: Bryn (Hill); Canol (Centre); Coed Cae (Field of Trees); Cog (Cuckoo); Daren (Rock); Eirw (Rippling, wavy stream); Fach (Small); Fawr (Big); Gwaered (Descent); Hafod (Summer dwelling); Heol (Road); Isaf (Lower); Llanwynno (Parish of Saint Gwynno).

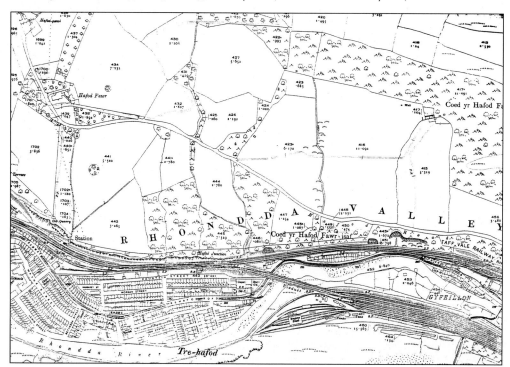

A Plan of Hafod in 1919. Place names: Llwyncelyn (Hollybush); Nythbran (Crows Nest); Pen-Y-Waun (Top of the Moor); Pistyll (Spring); Rhondda (Noisy, Babbling River); Tre(f) (Town); Tŷ Mawr (Big House). (Mervyn John)

Lewis Merthyr Colliery in 1905. Village Street Names: Rheolau Terrace, Woodfield Terrace, the first bathroom to be built in the village was at No.6 Mount Pleasant, Oaklands, Bryn Eirw, Pleasant View, Bryn Eirw Hill, Church Hill (Parish Hill), Coed Cae Road, Sant's Hill, Phillips Terrace, Temple Buildings, Fair Oak (a small hamlet), Hafod Side, Bryn Ffynon Cotts, Cwm George, Daren-Y-Pistyll Terrace, Bridge Street, Trehafod Road, Thomas Place, Cadwgan Terrace, Ael-Y-Bryn, Margaret Street, Cross Street; Morganwg Terrace (Morgan Street), Railway Terrace, Fountain Terrace, Wayne Street, Lewis Street, Bethesda Street, Ivor Street, Great Street; Western Street, Colliery Street, Afon Street (River Street).

On 1 February 1934 the last tram ran in the Rhondda and as it returned to the depot at the end of the day, a huge crowd sang 'Farewell My Own True Love'. The following list of names that were the head of the house and were amongst the first inhabitants of Trehafod: W. Arthur, W. Bassett, G. Daniel, J. Davies, M. Davies, W. Davies, B. Edwards, E. Edwards, M. Evans, W. Evans, W. Ford, J. Glanville, J. Griffiths, W. Hardwick, T. Herbert, J. Israel, J. John, J.D. Jones, T. Jones, T. Lewis, D. Llewellyn, J. Lougher, J. Macindo, J. Mathews, W. Mathews, W. Meredith, D. Morgan, E. Morgan, E. Morgan, J. Morgan, L. Morgan, M. Morgan, D. Morgans, E. Morgans, R. Phillips, W. Powell, B. Rees, M. Richard, W. Richard, J. Rimson, M. Rowland, W. Rutta, D. Thomas, E. Thoma, H. Thomas, J. Thomas, J. Thomas, W. Thomas, A. Watts, H. Wheeler, H. Williams, J. Williams, P. Williams, T. Williams, W. Williams. (Mervyn John)

Sant's Hill in 1965. Locally-known Sant's Hill was named after Mrs E. Sant who once owned the Vaughans Arms. It is typical of the size of houses that would house the miner, his wife, several children and a couple of lodgers. In 1891 the average house in the Rhondda was inhabited by 6.5 persons (compare to the industrial areas of England which averaged 4.3 to 4.7 at this date). The effect of subsidence is very marked. When one considers that the valley floor at Trehafod dropped 14ft in the time the mine was working it is amazing that anything in the village is left standing at all!

Sant's Hill in 1965. The number of shops that grew in the village evidently proved the self-sufficiency of the people in a rapidly expanding industrial area. On both sides of the main street the shops were built next door to each other, some trading from their front rooms. At one time there were nearly as many shops in Trehafod as in Porth or Pontypridd. Evans the Grocer, Morgan the Butcher and Thomas the Fruiterer were just a few of the businesses run in the village. Everything that was needed for the people was in existence – a cobbler, cafés, coach builders, midwife, dentist, post office, bakehouse, fish shops, undertakers, choirmasters and even musicians. Company stores were also used where goods could be bought in exchange for tokens. (K. John)

View of Lower Eirw Bridge, River Rhondda and the village in the 1960s. The River Rhondda is made up of two rivers, the River Rhondda Fawr and the Rhondda Fach, which meet in Porth. The River Rhondda then flows down the valley and joins the river Taff at Pontypridd. Before coalmining and the industrialisation of the Rhondda the river was very clean and a haven for fishermen. The coal industry came along, however, and the river became a black torrent of water – hardly a fish to be seen. Now that the mines have closed the river has returned to its former glory, with fish once again jumping in the river.

Lower Eirw Bridge with the Royal British Legion Club and Temple Buildings Trehafod in the 1960s. The Eirw Isaf Bridge was constructed to span the River Rhondda and has always been a vital link between Pontypridd and the Rhondda. Of steel construction, the bridge was widened in 1907 and, to accommodate the ever-increasing traffic, it was rebuilt in 1927. Still standing today, it has carried many thousands of all types of vehicle over the years, especially before the Trehafod bypass was built. With the outbreak of the Second World War the authorities were fearful that if the bridge was damaged the main link between the Rhondda and Pontypridd would be severed, consequently a narrow road was built between Trehafod and Llwyncelyn – not named but known locally as the 'Burmah Road'.

The Memorial Plaque on Lower Eirw Bridge. It was built of steel because stone bridges would collapse due to subsidence. The impressive ends of rather a small bridge were no doubt the council's policy of improving civic pride during the depression. But the bridge was made in Shropshire, not South Wales; like coal, the Welsh steel industry was in recession. The great majority of the newcomers who came to the mining settlements of Dinas, Cymmer, Hafod and Ynyshir, were agricultural labourers taking the change from farm labourer to miner. The new villagers occupied cottages owned by the colliery companies. During the first half of the nineteenth century the district was destitute of drainage arrangements with no provision of any kind for the disposal of excrement or refuse.

The waterfall at Daren-Y-Pistyll in 1910. The houses were demolished in 1959. Besides rainwater, the only source of fresh water was the mountain springs. The new villagers were self-sufficient, owning their own allotments, which could be found on either side of the valley mountains. These allotments supplemented the main source of food for the villagers and they took great pride in the fact that they provided their own living. Families at this time were large and for some it was a struggle to survive. Poverty in the Rhondda and Trehafod was widespread despite the wealth of the mines. The village was predominantly a male environment, where very few women could find employment. In the event of loss of husband, father or brother, many women were forced to leave their homes in search of work and migrated to parts of England, seeking employment mainly as domestic servants and mother's helps.

Trehafod Road in 1975. In the 1880s and 1890s tinsmith Howel Thomas hand made miner's drinking water jacks and food boxes (tommy boxes) and he also repaired saucepans (usually plugging the leak). Billy Powell's Garage (seen behind the car) sold motorcars, petrol, sweets, motorcar insurance and ran a taxi business.

King George V and Queen Mary at the Arch of Coal, Trehafod, Thursday 27 June 1912. In 1851 the neighbouring Gyfeillon had a population of seventy-six, and Gwaun-Yr-Eirw had a population of 175. By 1881 the population of Trehafod had grown to approximately 2,000, increasing in 1901 to 2,500 and reaching a high of about 2,750 in 1921. The population shown in the 1991 census was 978 people, (about three to a house instead of the six in 1851 and eight to a house in 1921). These figures approximately mirrored the Rhondda figures.

On the morning of 23 January 1911 a thunderous crash was heard coming from the Taff Vale railway line between Trehafod and neighbouring Hopkinstown. Those who may have feared the worst and rushed to the scene were horrified by what they found. Lying crumpled and smashed at the rear of a parked coal train was the 9.10 a.m. Treherbert-Cardiff passenger train. Eleven people died on that morning, with the guard of the coal train dying later in hospital.

Re-constructing the events leading to the tragedy, witnesses described how the passenger train had rounded the bend, oblivious to the presence of the stationary coal train. Travelling down a gradient at a speed of 30mph, the train's crew never stood a chance of avoiding the impact. Echoing more recent rail disasters of our time many passengers experienced miraculous escapes from the mangled wreckage.

Drafted in to help with the rescue operation were soldiers from the West Riding Regiment and officers of the Metropolitan Police. Ironically they were in the area to deal with possible strike action by Rhondda miners, a number of whom were on the train. They were part of a delegation headed for the Miners Federation of Great Britain conference in London. Amongst the dead were three Rhondda councillors and miners delegates, Thomas George, Thomas Harries and William Herbert Morgan.

The highly publicised inquest which followed was expected to lay the blame at someone's door. A confused coroner, Mr R.J. Rhys, sitting at a packed hearing in the New Inn, Pontypridd, recorded an open verdict. This decision reflected the layman's bafflement of the Taff Vale Railway Co.'s (TVR) complex signalling arrangements.

It was decided the failure by the coal train's crew to report the stoppage was a major factor in the incident. Mr Rhys stated the coal train fireman should have enacted rule 55 of TVR regulations, which involved alerting signalmen and detonating the line behind his train. But the inquest was told the fireman had been instructed by his driver to attend first to the engines lubricators. By the time the nearest signal box was finally alerted, it was all too late. General confusion that morning over which train was where meant that, to all intents and purposes, the coal train was 'invisible'. The situation was made worse by the equipment of the day, which was not suited to the tremendous intensity of traffic, placing unnecessary strain on the signalmen. The whole catastrophe highlighted a series of errors arising from outmoded and makeshift safety regulations and a remarkable laxity in their enforcement. Amid all the publicity, the Taff Vale Railway Co. was put under great pressure to improve its signalling radically.

As with many disasters, old and recent, it took a tragic incident to bring about much needed changes.

Trehafod Railway Station in 1998. The first railway to arrive in Trehafod was in 1835. Travel was difficult as pits were in existence long before this and miners found themselves trekking great distances to neighbouring pits. An everyday means of transport for the villagers was the bicycle and some boys would cycle for forty-fifty miles a night. Along with the railways one of the main forms of transport was the tram, which ran from Porth through Trehafod. The trams lasted until 1933, when the last tram passed through Trehafod. (K. John)

Trehafod Road in 1906. People migrated into Trehafod, primarily, for coalmining purposes and it is no surprise that the 1851 census lists: seventy-two coal miners; five labourers; three hauliers; three blacksmiths; two sinkers; one engineer and one coal agent. Other occupations that were listed include: mason; farmer; farm labourer; tailor; horse follower; and a railway policeman. Whilst the majority of women did not have occupations there were two dressmakers, two servants and a thirteen-year-old nurse.

Trehafod Road in 1906. The oldest working coal miner was seventy-two-year-old Morgan Morgan, whilst the youngest was nine-year-old George Morgan. The local magistrate was Lewis Morgan, who owned a considerable amount of land and lived with his family in Hafod Uchaf. It may be of interest to note that there were three widowers living in Trehafod. In the 1800s men tended to live longer than women.

Trehafod Road with the Trehafod Hotel in the background in 1906. The number of pubs also grew with the population increase in Trehafod. The Trehafod Hotel was first built as a Masonic lodge, with one of the rooms shaped like a three penny bit so there would be no corners for the devil to hide in. The Trehafod Hotel also played host to the Trehafod Band. The pubs would be full on a Sunday as it was the miners day off, a time for them to relax and wash away the coal dust.

Left: Pte T. Passey, First World War. By the time it was over in 1918, this war had involved nations from every continent, had killed and wounded more than 28 million people and had changed the balance of world power beyond all recognition. The First World War I began at midnight Tuesday 4 August 1914 and ended on 11 November 1918, with the official end on 28 June 1919 with the signing of the Treaty of Versailles, The Peace Imposed on Germany By the Allied Powers. *Right*: A-S W.J. Cooke, Second World War. This war also involved every continent, with massive battles raging on land, in the air and at sea.

The Drums of the 1st Battalion the Welch regiment in 1935. The photograph includes Robert George Venting, Drum Sergeant Major of Trehafod. The Second World War began at 11.00 a.m. Sunday 3 September 1939, when German troops attacked Poland. On 30 April 1945 Hitler committed suicide and on 8 May 1945 the war in Europe came to an end. On 14 August 1945 the Japanese government surrendered bringing the Second World War to an end. More than sixty countries took part in the War, over 55 million people died and it affected the lives of three-quarters of the world's population.

Putting a load of coal in the 'cwtch' in 1950. A load (delivery) of coal (one ton) was tipped outside the house of a miner as part of his wages, today coal is delivered in 56kg pre-packed bags, and is not allowed to be tipped on the streets. The haulier tipped the coal in Bridge Street, Trehafod as there was no access to Cwm George for a coal lorry and the coal had to be carried up the hill in old tin buckets to all the houses including Bryn Ffynon Cottages.

The Fireplace in 1950. The coal fire has always been a symbol of warm welcome in the valleys. The fire was also used for baking bread, cakes, cooking dinners, boiling kettles, boiling buckets of water for the bath and drying clothes. Before the pithead baths opened in 1952, the miner came home from the pit and bathed in a tin bath in front of the fire. The Fire Brigade was situated on the bridge by the Britannia Inn. The brigade would not be called to chimney fires; the doors in the house would be closed to stop the draught so the fire would eventually go out itself. There were also three policemen in the village, a sergeant who resided in the station at the bottom of Trehafod, a constable and a colliery sergeant. (Mervyn John)

A collier's welcome home from work in 1950. Trehafod village grew mainly as a result of the industrial revolution and as a result of the rich bituminous coal in the lower Rhondda and the accessibility of the Glamorgan Canal and purpose-built tram roads, the lower Rhondda saw rapid economic growth in the nineteenth century. With the demand for bituminous coal from London and Ireland came the supply of jobs.

Popular newspaper characters Andy Cap and Flo in 1977. With the industrial revolution came migrants into the Rhondda and Trehafod. People came from West Wales, Somerset, Gloucester and Shropshire. During the war there were also Jewish immigrants settling into the Rhondda, often great organisers building the light industries. One particular immigrant was a Russian who repaired panes of glass. He would go about his daily job by carrying glass on boards over his shoulders.

Christine Williams' fifth birthday party at Lewis Street in 1952. The photograph includes: Kathleen Thomas, Christine's grandfather, sister Janice, Christine Williams, Jennifer Ryan.

Ironmonger F.I. Morris in the snow of 1961.

The flood of 1979. Unfortunately the river, over the years, has brought misery to many households in Trehafod, breaking its banks and flooding the homes of many people, in particular the heavy floods of 1938, 1960 and 1979. As well as the devastation to homes the flooding claimed the life of a man. During the last few years the river has been dredged, new retaining walls have been built and the river slightly diverted. Hopefully the river will not breach the banks again.

Barry Sidings Countryside Park (The Lakes) in 2001. Janet Davies Mayor of Taff Ely officially opened Barry Sidings Visitor Centre on 24 July 1995. An area once scarred by South Wales great industrial past has been transformed into a countryside park full of surprises. Fine walks, duck ponds stocked with many varieties of fish, an abundance of wildlife, picnic and barbecue sites and an overall air of tranquility now warmly welcomes its visitors. The younger generation has also been well catered for with an exciting adventure playground and a professionally-designed BMX track. In 1884 the Barry Docks Bill was passed through Parliament and by 1889 coal was being transported directly to Barry from Barry Sidings, Trehafod. In 1890 approximately 2 million tons of coal were carried into Barry and by 1900 this had increased to 2.5 million tons of coal.

C & D's (Old Terry Stores) in Pleasant View (formerly called Fairview) in 2001. A typical day of the local shopkeeper in 1891 would begin with opening at 5.30 a.m., taking three shifts and closing at 11.30 p.m. A typical example was Davies Sweet Shop where thousands of people passed through to buy their chocolate and cigarettes. The shop was managed by the women of the family. Mr Davies was a builder and undertaker whose business started in 1890. The number of undertakers in the village and the surrounding areas reflects the high mortality rate before the existence of the NHS, and the dangerous colliery employment. (K. John)

Left: Bryn Eirw Street. Across the road from the Heritage Park a narrow road runs steeply uphill. As you reach the brow of the hill a surprise awaits. At the height of summer you are greeted with a blaze of colour and smiles of pride. Champions of the local 'In Bloom' competition for the last few years and proud overall winners of the 1999 Rhondda Cynon Taff award. House upon house is garnished with colour, a display huge enough to rank alongside any civic garden or flower show marquee. *Right*: Bryn Eirw Street. *Left to right*: Brioney Jones, Joel Davies, Katie Baldwin. Not only is the Colliery visible standing sentinel at the end of the street, but tiny model coal drams make ideal flower tubs in many of the gardens.

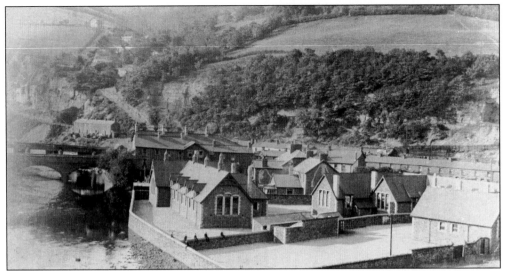

Hafod School in 1900. The schools consisted of three schools – mixed infants, boys and girls and were built in 1877. Much of the educational provision, in Wales, during the 1800s was through the astonishingly rich community life that existed in many areas. The choral festival, the Sunday school, the Eisteddfodau and the travelling schools. The Elementary Board Schools were set up under Forster's Act of Parliament in 1870. Under this act local schools were established and these included Graig, Coedpenmaen, Hawthorn, Mill Street, Navigation, Porth, Treforest and the Hafod school.

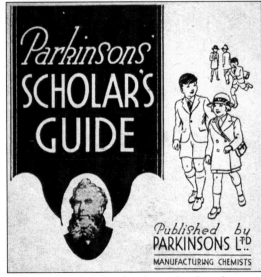

Left: Llanwonno School Board in the 1980s. The Hafod School was for the education of both boys and girls. Situated opposite the Vaughan's Arms, on Coed Cae Road, the master was Mr John Evans, the mistress was Elizabeth Thomas and the infant's teacher was Mary Evans. With the expansion of the local colliery and an increase in pupil numbers, the school was resited in Wayne Street and has remained there ever since. In the 1960s it became a primary school and retains the name of Hafod School. *Right*: Parkinson's Scholar's Guide. Thirty-four pages of educational information including; tables, weights and measures, English grammar, punctuation, abbreviations, history, geography, physical features, elocution, singing, etiquette, table manners, general behaviour, a page for girls, a page for boys.

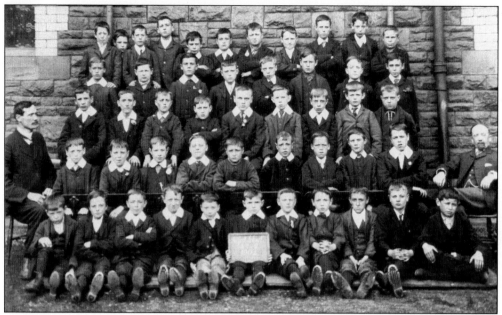

Hafod School Standard 5 and 6 in 1902. The photograph includes: Mr Evans, on the left, and pupils Daniel Davies and John Thomas. Like the population of Trehafod, the number of pupils has varied considerably. Probably the greatest numbers were 900, recorded in the 1940s, owing to the large number of evacuees that came into the area attending the school.

Hafod School woodwork class in 1902. Over the years the school has built for itself a fine reputation for good education. A recent Inspection by Government Inspectors considered the school to be one of the better Primary Schools in Rhondda Cynon Taff. This is due to the dedication of the staff, under the leadership of Mrs Jean Lewis, the willingness of the pupils and the support of the parents.

Hafod Junior School Standard 1 Christmas Nativity Play in 1953. *Left to Right, back row*: Christine Wilkins, Georgina Minty, Valerie Cooper, Averil Weekes. *Second row*: Gillian Bassett, Doreen Summers, Christine Williams, -?-, Merle Evans, Anne Hughes (Virgin Mary), -?-, Richard Williams (Joseph), Margaret Davies, Kaye Evans, Kay Guningham, Diane Minty, Carol Watkins. *Third row*: Kathleen Thomas; Jennifer Ryan, Ashley Cayton, Lilian Richards, John Bartlett, Michael Bird. *Front row*: Ceri Jones, David Rees, Brian Woods, -?-, Wesley Hill.

Hafod Junior School Standard 1 in 1954. *Left to right, back row*: Howard Pullen; Alan Phelps, Ceri Jones, Keith Davies, Melvin Jarman, Wesley Hill, Michael Bird. *Middle row*: John Bartlett, John Jones, Christine Wilkins, Margaret Davies, Kay Guningham, Susan Morgan, Richard Williams, -?-. *Front row*: Carol Watkins, Averil Weekes, Ann Hughes, Lilian Richards, Kay Evans, Kathleen Thomas, Georgina Minty, Pat -?-, Merle Evans, Gillian Bassett, Jennifer Ryan.

Hafod School on St David's Day, 1987. *Left to right, back row*: R. Richards, A. Moore, J. Hill, V. Lewis, C. Gooch, C. Pooley, S. Evans, M. Phelps, M. Shore, L. Hamilton, R. Robertson, M. Richards, S. Ashford, S. Lendrum, M. Emery, P. Williams, K. Jenkins. *Second row*: S. Davies, J. Markey, J. Emery, -?-, C. Howells, M. Parry, B. Snook, S. Waits, C. Pearce, -?-, K. Llewellyn, C. Waits, T. Shore, -?-, S. Stevens, -?-, L. Hughes, S. Waits, N. Cambridge. *Third row*: G. Williams-Jones, T. Pye, D. Shore, S. Howell, H. Llewellyn, -?-, C. Rees, A. Jones, R. Hughes, E. Lendrum, S. Ho, A. Pugh, -?-, G. John, G. Hughes, S. Williams, N. Howells, -?-, D. Lewis, D. Christopher, C. Pritchard. *Fourth row*: C. Trivett, A. Stevens, D. Helyar, S. Bailey, A. Solly, R. Kelly, J. Phelps, D. Rees, D. Fisher, -?-, C. Solly, M. Owen, C. Brocklebank, M. Button, A. Snook, D. Barrett, A. de la V. Zabala, J. Bowkett. *Front row*: S. Waits, R. Rosser, C. Williams, -?-, -?-, N. Hughes, A. Lewis, G. Howells, I. Hughes, S. Markey, L. Howells, L. Jones, C. Pugh, S. Brock, V. Davies.

Hafod School in the new millennium, 2001. Trehafod Village grew up dependent on the Lewis Merthyr Colliery. In 1956 a terrible explosion at the colliery wrenched the hearts of the community as nine men lost their lives as a result of the accident, leaving five others injured. Finally in 1983 the Lewis Merthyr Colliery closed, severing Trehafod's links with the mining industry and a way of life, if not spirit, which had been predominant for well over a hundred years. The area now boasts lush, green picturesque landscapes, which are a delight to see and enjoy. These age-old occupations have left the positive legacy of strong friendly community spirit, distinct character and a host of other traditions for which the village and the surrounding area has now became famous.

Three

Rhondda Heritage Park in the South Wales Coalfield

Hwy Clod Na Golud: Fame Outlasts Wealth. Such is the brave legend of the Rhondda Valleys and it can well serve to introduce this Park and its purposes.

For over 150 years the Rhondda and the other valleys that radiate from their hub at Pontypridd were an industrial dynamo transforming Wales. After the age of iron-making around Merthyr and the heads of the Valleys, came the intensive mining of coal, the frantic search for Black Gold that honeycombed the Glamorgan Hills and settled them with the new communities we now know as 'The Valleys'. Their fame was as global as the coal they gave to the world in an endless stream. The oceans were criss-crossed by the users and carriers of best South Wales steam coal.

Fortunes were made, expectations were turned upside down generation by generation, boom gave way to depression. There was despair and elation in equal measure. Through all the struggle a resilient people survived.

Now, after 1990, no more coal is produced in the very epicentre of that world, the Rhondda itself. So the Rhondda Heritage Park has been established on the site of the former Lewis Merthyr Colliery to present the story of Black Gold and the human civilisation that cocooned the mineral in its community culture.

The Rhondda Heritage Park will not preserve a dead theme in aspic, it will aspire to inform and inspire the twenty-first century with the meaning of the world that gave birth to the Rhondda Heritage Park.

By the beginning of June the hills were bulging with a clearer loveliness than they had ever known before. No smoke rose from the great chimneys to write messages in the sky that puzzled and saddened the minds of the young... The parade of nailed boots on the pavements at dawn fell silent. Day after glorious day came up over the hills that had been restored by a quirk of social conflict to the calm they had lost a hundred years before.

Gwyn Thomas

Left and right: The Colliery Abandoned in 1984. In 1983 Lewis Merthyr Colliery closed, should the buildings be demolished and the site become an industrial estate, or should the site be preserved as a 'tribute' to the mining history and the people of the Rhondda? A determined body of local people joined forces to ensure that their heritage was preserved for future generations. The initial feasibility studies began in 1984-1985, on behalf of a working party comprising Mid Glamorgan County Council, Rhondda Borough Council, Taff Ely Borough Council, the Welsh Development Agency and the Wales Tourist Board.

Discussing the future of the colliery. The photograph includes: Mattie E. Collins, MBE, Leader of Rhondda Council, and Edith May Evans, Mayor of Rhondda. In the early years of the establishment of the Rhondda Heritage Park substantial work had to be done. This involved extensive landscaping, including making six shafts safe, and generally transforming the site from one of dereliction and decay, to a first class amenity.

Rebuilding the perimeter wall in 1986. The Development Plan Report in 1986 was given by Consultants, William Gillespie & Partners, Leisure & Recreation consultants, Ove Arup & Partners, PEIDA, I.E. Symonds & Partners. The Rhondda Heritage Park will be a living testament to the coal-mining communities of the Rhondda.

HRH Prince Charles at Lewis Merthyr Colliery shaking hands with Ivor England in June 1987. The photograph includes: Mr M.A. McLaggan (Lord Lieutenant of Glamorgan), HRH Prince Charles (Prince of Wales), Ivor England (Tour Guide). Work b egan on site in 1987-1989. The Consultants were, Mid Glamorgan Land Reclamation Unit, Gillespies, Wyn Thomas & Partners and Bingham Hall O'Hanlon. Many original artefacts will be collected in readiness for the new centre.

Left: The Lewis Merthyr colliery hooter in 1987. The blare and wail of the colliery hooter sounded for miles along the narrow valleys of the Rhondda and was once a part of every day life. *Right*: Making safe the colliery stack in 1987. The design of Visitor Experience began at Lewis Merthyr Colliery in 1987. The Consultants were John Brown & Co., Wyn Thomas & Partners, Sally Wright & Partners, Heritage Projects Ltd and Frank Atkinson.

Left: The Colliery underground work in progress in 1987. The temporary Visitor Centre opens in refurbished colliery stores building in 1987. *Right*: Making safe the colliery headgear in 1987. 'Black Gold' is planned in 1987 using the Trefor and Bertie Engine Houses, the Fan Engine House and the Stores Building.

The dismantling of the Lillie Controller from Abercynon Colliery winding engine, which was given to repair the Trefor winding engine in 1987. The winding engine is fully restored and the maximum rope speed for men is 13ft per second. *Left to right*: Norman Creese (Electrician), -?-, Kerrigan Jones (Mechanic Class One). In July 1989 the Visitor Centre opened the stores building of the Lewis Merthyr Colliery to the public, providing a unique setting for the temporary exhibition galleries, gift shop and restaurant. When it opened it was only introduced to provide a taste of things to come for the local and regional communities, but as more than fifty per cent of its visitors came from outside Wales, the Visitor Centre became a permanent feature at the Park.

The Trefor Winding Engine in 1988. During 1989 and 1990, work continued on the Lewis Merthyr site to create the first stage of 'Black Gold – the Story of Coal', in the refurbished and restored pithead buildings. This opened in May 1991 and attracted over 35,000 visitors in its first year of operation. 'Black Gold' uses state of the art audio-visual and exhibition techniques to tell the story of people who lived and worked in the Rhondda.

The Lewis Merthyr Colliery pithead baths were used for storing furniture, 25 February 1988. Before the pithead baths opened in 1952, the miners came home from the pit and bathed in a tin bath in front of the fire. In the warm summers of yesteryear some night shift colliers would sleep on the side of the mountain taking advantage of the fresh air.

Left: The platform entrance to Hafod No.2 shaft, adjacent to Burmah Road, on 25 February 1988. First Pithead Festival at Rhondda Heritage Park opened to the public on 5 July 1988. *Right*: The Rhondda Heritage Park in the early morning light in 1988. During the first Pithead Festival on 7 July 1988 the winding engine house opened to the public.

The World of the Rhondda Heritage Park

The Western Mail informed its readers in 1904 that Pontypridd is not an earthly paradise:

> *It has not the elegant boulevards of Paris, nor the beauteous avenues of some of our English cities. A Cardiff lady took her little daughter to the capital of the Rhondda the other day ... and the child was obviously much impressed... After much cogitation in her childish mind, she asked her mother, 'Mamma, is this one of the places where the wicked people are sent to live?'*

The joke, in reality, backfires on Cardiff. That city, the future capital city of Wales, was itself described in 1905 by a French visitor, awestruck at the industrial energy and urban chaos of Coalopolis, as a 'Welsh Chicago'. Cardiff, and all that touched upon it, owed its pre-eminence and its growing pretensions to civic beauty to the wealth generated by the digging of coal in the valleys of the Cynon, the Taff, the Ely and the Rhondda. The main rivers mingle together at the confluence of Pontypridd and, at the beginning of this century through Pontypridd, by day and by night, hour after hour, coal-laden trains from Rhondda and the adjacent valleys steamed and clattered to the voracious docks at Cardiff and Barry where their Black Gold was shipped out to fuel the whole world.

Pontypridd – first a fording point where rivers join, then a small market town graced since 1756 by the elegant bow of William Edwards' single-span bridge over the Taff, became, by the 1850s, a scatter of houses and small works conveniently labelled 'Newbridge'. It was transformed by the twentieth century into a pulsating town of over 40,000 people, further swollen by outlying villages and collieries at Cilfynydd, Beddau, Ynysybwl and around the ancient settlement of Llantrisant, sentinel to the rural past of the Vale of Glamorgan and of the modern valleys to the north.

This locale was the very eye of the storm that was South Wales where in 1911 two out of every three people in Wales had come to live.

And still it did not stop – in the early 1920s the two narrow Rhondda Valleys held nearly 170,000 people alone, brought there to contribute their manual efforts to produce nearly 10 million tons of all the Welsh coal that had peaked at a tonnage of 57 million in 1913. It was on the back of this that Cardiff and the newer docks at Barry then exported more coal through their harbours than any port has ever done in history.

You can still sense such power by wandering up and down the ever-curving railway platform on Pontypridd's magnificent station, an architectonic hymn to iron and stone whose monumentality testifies to the dynamic force that made the town and its hinterland erupt.

This was the world Black Gold made and still to be seen all around are the traces of the people who first swarmed here to work then stayed to live and create these still lively communities. The landscape, of sheer mountain and plunging 'cwms', is startling enough in itself but combined with the basic needs of the people, it has been re-shaped into stunning vistas that draw the eye up nigh perpendicular streets or lay out a tumbling fantasy of roofscapes.

Along twisting main roads, hemmed in by railways and rivers, Victorian chapels in grey dressed-stone or Edwardian hotels and public houses in bright red-brick suddenly loom in a prideful, competitive bulkiness whilst the hills, though their plateaux remain wild and enticing, are claimed for habitation by the townships which have ridged and lined them to their very tops with the crenellated battlements of terraces built to house the people.

The sense of packed-in power and pent-up vitality stems from the rapid eagerness with which it was all accomplished. In the half century to 1914, the shifts in people's expectation of what was normal or settled altered with bewildering regularity.

In Newbridge, which was turning into Pontypridd, two weavers, the father and son, Evan and James James, sat down in Mill Street to write the words and music of their Welsh song

'Hen Wlad Fy Nhadau' (Land of My Fathers). That was in 1856. No one took much notice for a while. Forty years on, as the newly formed Welsh Rugby XV began to win at Cardiff Arms Park on the banks of the River Taff with a metronomic certainty bred of widespread social confidence, the watching crowds began to sing the air with such fervour that, soon, no one doubted it was, indeed the Welsh National Anthem.

Those same expectant spectators were also cheering a sporting phenomenon, which the Press labelled 'The Rhondda Forward' to denote the pit-hardened muscle from the coalfield, which Wales now carried in her pack. Behind the rock-steady scrum flitted jinking, swerving outside halves like Rhondda's Percy Bush, the prototype for Rhondda's Cliff Jones in the 1930s and Cliff Morgan in the 1950s, and scurrying wingers like Teddy Morgan from Aberdare and Willie Llewellyn from Llwynypia.

All three played in the victorious Welsh team against the otherwise invincible New Zealanders in December 1905, Black Gold beat the All Blacks. Percy Bush became Cardiff's Consul in Nantes whose trade links in coal and pine wood for pit props is remembered to this day in Cardiff's Boulevard de Nantes, Willie Llewellyn became a chemist in Tonypandy and had his shop left unharmed when a strike turned into a riot in 1910 and Winston Churchill sent in troops to protect the other shopowners, and Teddy Morgan who scored the only and winning try that day was regaled by the exultant crowd with a new hymn tune composed in 1905 by John Hughes, a colliery clerk form Llantwit Fardre, near Pontypridd. Those who would from then on call upon Jehovah to guide them and Wales to win did so to the music of 'Cwm Rhondda'.

By the turn of the twentieth century this was a place name already automatically coupled with the steam coal industry of Wales all over the world. The fame echoed in unexpected ways. It was carried to Canada by Rhondda coalowner, D.A. Thomas who named a peak in the Rockies 'Mount Rhondda'. He, himself, went to the House of Lords in 1917 as Lord Rhondda, a year after a collier boy from Tylorstown who never fought over 7 stone in weight began his illustrious reign as flyweight champion of the world.

Jimmy Wilde's great hero was the first Welshman to become a world boxing champion in 1914. That was Freddie Welsh of Pontypridd who lost his lightweight title in New York City in 1917. Twenty years later, in 1937, when Tommy Farr from Tonypandy entered the arena in that metropolis to face Joe Louis and gallant defeat over fifteen unforgettable rounds, the robe draped over his back, yellow silk emblazoned with a red dragon, was that of Freddie Welsh and a vanished world.

The Valleys had been battered out of their self-confident swagger by the 1930s. The end of the First World War in 1918 had seen the scythe of economic recession taken to the pits and the population of South Wales. Yet, oddly, the renown of Rhondda and the Valleys sounded even more in depression years between the Wars. Perhaps the outside world recognised fully the drama that had unfolded. There was a hungry fascination to uncover the meaning of this incredible story. Richard Llewellyn in his 1939 international best seller, *How Green was My Valley*, and John Ford in his oscar-winning film of 1940, fed the appetite with romance. The world is still hungry to know.

The purpose of Rhondda Heritage Park, as the era of Black Gold finally ends, is to fulfil that need by presenting and projecting that story, in all its varied aspects, to as wide an audience as possible.

Our story has been told in this way by our own writers. It is there in the sly, elegant short stories of Clydach Vale's Rhys Davies (1900-1978) and in the post-war revelations of the hilarious fiction of Pontypridd's Alun Richards (1929-). Lewis Jones (1879-1939) wrote more authentic epics in his panoramic novels than Richard Llewellyn ever managed and novelist Ron Berry (1920-) brought a scalpel from Blaencwm to shave the excesses off the epic whilst in Porth's Gwyn Thomas (1913-1981) Rhondda produced a writer of black, satiric fables who stands comparison with the world's best.

Painters, too, have re-interpreted shapes and colours to give extra dimension to the black and white images of film and photograph.

Actors, from Donald and Glyn Houston to Stanley baker and singers, from Sir Geraint Evans and Stuart Burrows to Tom Jones, have all informed their public and professional lives with the individual and communal spirit that these places gave them.

The Rhondda Heritage Park is a homage to all that was created here, from the ordinary heroism of daily survival to the extraordinary act of outstanding people throughout these astonishing communities. The Park, rooted in the Rhondda, is a branch of the line connected through Pontypridd to all the proud and distinctive history of Mid Glamorgan. And, as the history of Black Gold now has, perforce, to be imagined for future generations the Rhondda Heritage Park will be the key that unlocks the memory of that future.

Dai Smith

Secretary of State Peter Walker officially opened Rhondda Heritage Park on 11 September 1989. Housing a fine exhibition of valley life, an art gallery coffee shop and quality gift shop. Development commenced in 1987 after the then Secretary of State for Wales, the Right Honourable Peter Walker MP, included the Park in his three-year programme for tackling the problems, which existed in the South Wales Valleys – 'A Programme for the People'. In that programme the Welsh Office committed approximately £2.2 million to the project and this was matched by the organising partnership producing a capital development budget of just over £4.4 million to create the first phase of the overall scheme which in 1986 had been estimated at around £9 million. Other funds have been received from the Countryside Council, the European Community, Industry and Business.

The Rhondda Heritage Park in 1990. The second phase of the project opened in 1991 with a fascinating array of audiovisual presentations and special effects, transforming the Trefor and Bertie winding houses, with their engines, into huge industrial theatres providing an illuminating insight into the lives of miners and their families.

Two Trehafod schoolboys dressed as colliery boys at the Rhondda Heritage Park in 1990. On 9 October 1990 spectacular scenes were planned for tourists. The story of coal, as it has been related through South Wales mining villages for decades, should be greeting tourists from next summer. That is when Rhondda Heritage Park hopes to have finished constructing the surface section of its Black Gold Experience. Using the refurbished pit buildings, including the giant winding gears of the Trefor and Bertie shafts, the above ground show will tell the 140-year history of mining in the valleys. Drawing on the personal experience of many who worked down the pits, the story will reconstruct the lives of three generations of miners from the nearbye village of Trehafod who all worked at the Lewis Merthyr Colliery around which the Heritage Park is planned. Grandfather George Rees started down the pit at the age of twelve.

Left: Tour Guide Ivor England taking school children on a guided tour of the colliery in 1990.
Right: Christmas Advertising Brochure in 1990. Rhondda Christmas Celebrations at the Rhondda Heritage Park, Saturday and Sunday 15 and 16 December 1990.

Christmas celebrations at the Rhondda Heritage Park, December 1990. Father Christmas (Siôn Corn) is in his underground grotto giving presents to the girls and boys. Using archive material and written personal accounts the research team has recreated what a miner would have seen, including the arrival of the Taff Vale Railway, which opened up the Rhondda valleys for trade. The character describes the gold rush mentality, which brought thousands flocking to the area, doubling the populating in just ten years. His son John then takes up the narrative, telling visitors how dangerous mining in Rhondda was and describing operations carried out on injured men in front parlours and even on the undertaker's slab.

Left: The making of model miners for the Rhondda Heritage Park in early 1990. By spotlighting various areas of each of the buildings, the creative team have fitted several tableaux into one, showing different stages of history. Film and video techniques, with historic scenes and cuttings played back on specially situated screens, will add to the impact. *Right*: Model miners on display in 1990. The terror of an explosion underground will be illustrated by showing frantic scenes on the screens accompanied by authentic rumblings while the visitors are plunge into pitch darkness. Buildings have been renovated, including the former lamp room and the Trefor and Bertie engine houses.

Friends of the Rhondda Heritage c.1990. The photograph includes: Tim Richards, Howell Williams, Shirley James, Ron Clarke, Tegwen Waygood, Pearl Penrose, Jenny Tann, Jim Squires, Betty Hopes, John Gair, Ray Hopes, Martin Doe, Phil Price, Alf Carpenter, Brian Morris, Sir Donald Walters, Colin Jones, Lynne Harris, Mike Nash. The concept and detailed design of 'Trefor and Bertie's Energy Zone' began in 1992, an interactive children's play area on the upper colliery yard. The consultants were: Richard Ellis, John Sunderland, Mid Glamorgan County Council Landscape Architects, Geoff Cheason, and Architect, Ian Parfitt. The Marketing and Business Plan for the Park's further development began in 1992. The consultants were Coopers and Lybrand Deloitte, Leisure and Recreation Consultants.

Left: Underground Adventure. The Visitor Centre has been extended to double its size and the new simulated 'Underground Experience' allows the public to be lowered in a mine cage as if descending the 1,500ft to the pit bottom. *Right*: Discover the secrets of an underground world, go back through time and unlock the secret, sights and sounds, tragedies and triumphs of a world known only to a brave band of men – Welsh miners.

Bertie the Canary having fun on Trefor the pit pony's head. The amazing special effects in the Underground Adventure at the Rhondda Heritage Park make it all seem so real. There are lots more besides that make it an interesting and fun day out for all the family: from the Black Gold Trail, which shows just how life was for mining folk 100 years ago, to the exciting hands-on exhibits and play activities for kids in Trefor and Bertie's Energy Zone.

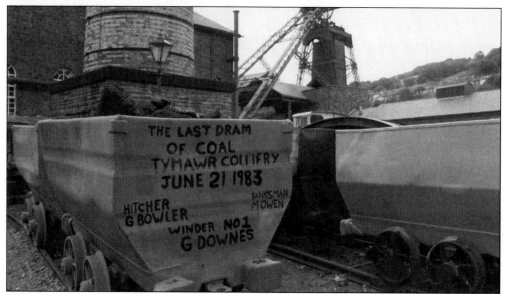

The last dram of coal raised at Tŷ Mawr Colliery, 21 June 1983, on display at the Rhondda Heritage Park. The Rhondda Heritage Park would to like to thank Bryn Rees who was an inspiration for the story of 'Black Gold'. Bryn was an everyman chosen to tell the tale of a vanished era that should not be allowed to fade from memory. Sadly on 30 January 1991 only days after his eightieth birthday and only weeks before the realisation of a dream, Bryn died. 'Black Gold' is a tribute to Bryn and thousands like him.

Garry Townsend (third from left) with a group of visitors in 1990. Mardy Colliery minecar: the last dram of coal raised in the Rhondda Valley on 30 June 1986 on display at the Rhondda Heritage Park. Since the beginning of coal production at Mardy No.3 and No.4 the mine has extracted 12.8 million tons of coal and developed 165 miles of underground roadways, equalling the distance from Cardiff to London. In the 1960s the colliery employed nearly 2,000 miners.

Singing superstar Frankie Vaughan visits the Park in 1990. *Left to right*: Clive Thomas (former International and World Cup Football Referee), Andrew Hood (Heritage Park Director), Frankie Vaughan. The Friends of the Heritage Park are people from all walks of life, who live both locally and further afield. They have a common aim of supporting this exciting project from the beginning of its development to the stage where it is now one of Britain's major tourist attractions.

Tree Planting, Christmas 1991. *Left to right*: Alf Carpenter (Friends), Pat Carpenter (Friends), Mair Morris (Friends), Brian Morris (Friends), Andrew Hood (Director), Collin Jones (Friends). The Friends provide voluntary help in many ways – stewarding during the Park's special events or undertaking voluntary duties around the Park. They attend various shows and events and run heir own projects through an independent committee to help promote the Park.

Mr David Hopkins, the village cobbler, in 1995. Hopkins was born in Pentre Road, Maerdy, in May 1906. When the building of Richard Street was completed in 1911 his family moved to No.4, where he lived for the rest of his life. In 1920, at the age of fourteen, he commenced his apprenticeship as a cobbler with Mr Gibbons who had a shoe shop and repairing business in Ceridwen Street. He worked there until 1930 and then started his own business at 4 Richard Street. His customers were generally miners and their families.

The Cobbler Shop, 24 August 1998. *Left to right*: Jill Morgan, Jennifer Griffith, Mary Lockyer. Mr Hopkins worked in his cobbler shop until 1996. He gave up work at this time, at the age of ninety, due to failing eyesight. He enjoyed two years retirement and died just before his ninety-second birthday. He was respected by those who knew him and died following a fall at his home. This 'unique character' will be greatly missed by many people. Mr Hopkins' cobbler shop was dismantled and taken to the Rhondda Heritage Park and is waiting for a suitable space where it will be permanently on display.

Glyn Houston on 13 May 2000. Actor Glyn Houston unveiled a memorial to the thousands of miners who died in the South Wales Collieries. The memorial is in the form of a six-foot-high miner's lamp, complete with an eternal flame, which will stand as a lasting tribute to all the miners who died while working in the pits. In the early years of the establishment of the Heritage Park, substantial work had to be done. This involved extensive landscaping including making six shafts safe and generally transforming the site from one of dereliction and decay to a first class amenity.

Poet and songwriter Hawys Glyn James read out a poem she had composed for the unveiling of the Coalfield Memorial of a miner's lamp on 13 May 2000. Friends are entitled to free admission to the Park when accompanying paying visitors. They all feel personally involved in the success of the project and keep in touch through regular newsletters. They organise many activities including special events, visits to places of interest and talks from visiting speakers.

The Miner's Lamp prior the unveiling on 13 May 2000. The man-size lamp, set on a stone plinth, is an exact replica of the working lamps manufactured in the Cynon Valley by E. Thomas & Williams Cambrian Lampworks. One of the Cynon Valley's oldest established businesses, which still supplies authentic miners' lamps worldwide, the company is based in Robertstown. It traces its origins back as far as 1860. 'We feel honoured that our company has been recognised in this way,' said a company spokesman, 'E. Thomas & Williams Cambrian Lampworks is synonymous with the coal industry and we are proud to be featured in such a way.'

Friends of the Heritage Park in the new millennium. *Left to right, back row*: Jill Morgan, Joyce Jones, Brian Morris, David Owen, Haydn Shadbolt, David Morgan. *Front row*: Vera Lewis, Kathleen Thomas. Attending the unveiling of the memorial were: Mr David Morgan (Chairman), Sir Donald Walters (President), Rita Moses (Chairperson, Rhondda Cynon Taff County Borough Council), Revd Malcolm Chiplin, Vicar of Pwllgwaun, Lewis Merthyr Band, local MPs Kim Howells and Allan Rogers. After unveiling the memorial, Rhondda-born actor Glyn Houston said, 'I feel very privileged to have been asked to unveil this very fitting memorial. My family had close links with the mining industry and this is a wonderful tribute which I am sure will be cherished.'

Left: First Bond of the shift for the Rhondda Heritage Tour Guides in the new millennium. *Left to right*: Ivor England, Graham Williams, Alun Davies, Garry 'Busty' Mason, Peter Arundell. Fourth from left, Busty, was employed at the South Wales Collieries all his working life. In 1957, at the age of fifteen, he started work at Tylorstown No.9, he also worked at Mardy, Tŷ Mawr, Nantgarw and Trelewis mines. *Right*: Underground with tour guide Busty in the new millennium. The colliery, which he worked at the longest, was Tŷ Mawr/Lewis Merthyr. His hobby has always been showing dogs and he first started showing Afghan Hounds in 1970. Busty is a Championship Show Judge.

End of the shift for the tour guides Alun Davies, Ivor England, Peter Arundell, Terry Conway, Graham Williams and Howard Lewis in the new millennium.

Tŷ Mawr was a hard mine to work. We were still filling coal by hand, with a size four shovel until 1976, but we had some fun from the time we arrived in the baths until we got to the coalface. We were laughing at something, or what had happened to someone. The mine was full of characters. Jack Evans was one of these or 'Shoni Mount' as he was known. Working on the same face together as youngsters we would get the coal off and timber up later. Shoni shouts up the face, 'Tell the boys to timber up. There are three big bosses coming up the gate road.' I said, 'Who are they Shoni?' He said, 'Was 'is name, "Watucallhim" and I'm not sure of the other one.' On one occasion there was a fall at the coalface and the men were sent to other districts. I stayed one end of the fall with John Peal and Vernon Davies and Shoni was working the other end. The fireman was John Hughes who had a stammer. The phone rings my end so I answer and a voice says, 'How how how are things looking that end of the fall?' I ask, 'Who is this speaking?' The voice says, 'This is the the the un un un undermanager here'. So I say, 'We have put up three pairs of tim-ber this end, John,' he said, 'How how how did you know it was me, Busty?' I said, 'I just guessed John.'

Tour guide Alun Davies gives a talk to school children in front of the Period Village Street, with the contents of an entire butcher's shop and general stores, retaining many original features, as part of the Rhondda Heritage Park's living mining village display in 2001. In 1991 the Rhondda Grocer's Association gave the balance of their funds to help Rhondda Heritage Park collect and keep the shop fronts from High Street, Gilfach Goch.

Director Mick Payne handed over the keys to new Director John Harrison on 12 February 2001. During the years of the Rhondda Heritage Park there have been four directors: Andrew Hood, Catherine Hill, Mick Payne, and John Harrison.

Work begins on the construction of Energy Zone and support building in 1992. The contractors were Monkey Business, Farmer Studios, Gerald Davies Ltd, Richard Glassborrow and Edward Davies Ltd. The consultants for the 1993-4 detailed design for the Underground Experience and site interpretation were: Brooke & Associates, the Visual Connection, Roger Richards Partnership, Bradley & Associates, Haley Sharpe Associates, Harley Haddow Partnership, and the Heritage Park guides.

Building the children's Energy Zone in 1992. In the background is the Rhondda Heritage Hotel and former Coed Cae Pit site. The hotel opened in August 1991 and held its 10th Anniversary Celebrations on 25 August 2001. Set in the famous and beautiful Rhondda Valley the hotel is an ideal base from which to explore South Wales, the Brecon Beacons and Cardiff. It is owned by Peter and Gill Hands and has forty-four modern en suite bedrooms, conference and banqueting facilities for up to 200 guests, and a fully equipped leisure club with indoor pool.

Left: Fun in the Energy Zone. In May 1993, the Energy Zone opened, where children can learn about coal formation. *Right*: Join Trefor and Bertie in the Energy Zone for a fun, action-packed adventure – climb the Temple of the Sun, mine your own energy in Castle Coal and discover how coal was made.

Exploring the energy cycle in a unique, fun and educational way, the Energy Zone provides a new dimension at the Park. Bertie and Trefor celebrate their first birthday with school children in the Energy Zone.

Dr Barnardo's children named the Rhondda double-decker bus *Bertie* on 31 July 2001. A birthday party with a difference was held at the Rhondda Heritage Park on 31 July 2001 when staff celebrated the 40th birthday of a double-decker bus. The double-decker, which is just one of the tourist attractions at the centre, was built by the Metropolitan Scammel Carriage & Wagon Co. and first registered in 1961 with the Rhondda transport Co. The AEC regent bus was one of ten owned by the Rhondda Co. and was operational for ten years providing a service between the Rhondda Valleys and Cardiff.

Rhondda Transport bus ticket was issued by conductress Jill Morgan on 31 July 2001. The bus was manned by a driver and conductor before gradually changing to a front-entry, one-man operated vehicle. The fleet of double-deckers was eventually sold to an operator in the West Country before the one, which is now at the centre, was purchased and returned to the Rhondda in 1991. Through the Friends of the Rhondda Heritage Park the bus is regularly used for many events including fetes and fairs.

The happy fortieth birthday of a double-decker bus proved to be a 'moving' occasion attended by local children and youngsters from Dr Barnardos. 1993-1994 saw the construction of the new Visitor Centre and Village Street, contractors: William Cowlin & Sons Ltd, Steel Design Services, Dolcast Engineering, MITIE, Flair Electrical, Haywood and Williams, Stage Electrics, Lift Services Wales Ltd, Jane Linz Roberts, Jarnie Garvin and the others.

Face painting on 30 July 2001. The development, the new underground tour, 'A Shift in Time' opened to the public in July 1994. A Shift in Time includes a cage ride to 'Pit Bottom', guided tour through the underground roadways of the Lewis Merthyr Colliery pre-mechanisation and the working coalface. To complete the underground tour there is an exiting finale – a mysterious and unforgettable route back to the surface.

On a snowy Christmas Eve. *Left to right*: Daniel Davies, Father Christmas (Siôn Corn), Zoë Davies. In 1999 a Visual Exhibition opened with a 500sq. ft mural depicting the Tynewydd Colliery Disaster of 1877. The Tynewydd Colliery was sunk in 1852 at a depth of 270ft by the Troedyrhiw Coal Co. At around 4.00 p.m. on Wednesday 11 April 1877, the Tynewydd Colliery was inundated with water from the old workings of the adjoining Hinde's Upper Cymmer Colliery. At the time of the inundation there were fourteen men in the pit, of whom four were unfortunately drowned and one killed by compressed air, leaving nine men imprisoned by the water, of this number four were released after eighteen hours imprisonment and five after nine days imprison-ment. It was in effecting the release of these latter five that those distinguished services were rendered which the conferring of the Albert Medal of the First Class is intended to recognise. Then came Queen Victoria's announcement: 'The Albert Medal, hitherto only bestowed for gallantry in saving life at sea, shall be extended to similar actions on land and that the first medals struck for this purpose shall be conferred on the heroic rescuers of the Welsh Miners.'

Season's Greetings from Rhondda Heritage Park

Cyfarchion yr Ŵyl oddi wrth Parc Treftadaeth Cwm Rhondda

The Rhondda Heritage Park domestic staff ('mopathologists!') in the new millennium. *Left to right*: Chris Howells, Cilla Thomas. Our miners, workers, our choirs, our scholars, our actors, our writers, our sportsmen, our movements of protest and reform, our 'characters', our passion for the community, all of it exists as proudly and as positively today as it has done for almost 200 years.

The Rhondda Heritage Park Staff. *Left to right*: Nicola Newhams, Lisa Burnell, Mari Davies, Janet Pennell, Jemma Simons, Jennifer Griffith. In the spirit of the region, a friendly local staff contributes towards the informal atmosphere, which pervades the Rhondda Heritage Park and ensures that every visitor is given the warmest welcome and are keen to help people with special needs, even underground. Purpose-built paths for wheelchair and pushchair users provide easy access.

The Rhondda Heritage Park Staff. *Left to right*: Sarah Tovey, Emma Davies, Linsey Morgans, Anna Tovey, Gillian Lewis, Roger Griffiths, Bethan Samuel, Robert Summerhill. If you would like to experience this welcome for yourself and discover the charm of the South Wales Valleys, please do not hesitate to contact us.

126

Lewis Merthyr Colliery, the Rhondda Heritage Park in the new millennium with Ross and Luke Wilding. Preparing to settle down to a well-earned sleep where Rhondda people are renowned for their warm and friendly welcome. Here nature has returned and has carved a beautiful backdrop of rolling hills, vales, mountains and moorland. It was the mining of coal, on a massive and savage scale, that brought Rhondda's fame and, for some, fortune. But the people who chose the Borough's motto – *Hwy Clod Na Golud*, (Fame Outlasts Wealth), chose well, for the mining of coal, and the wealth that went with it, has now gone but Rhondda's fame lives on and, indeed will gain a measure of immortality with the exciting and imaginative Rhondda Heritage Park.

The Valleys are transformed from being the centre of heavy engineering and mining, to being green and lush. They retain their sense of pride and traditional heritage but combine it with a new optimism. Never has the phrase 'How green is my valley' been more appropriate.

A Proud Past Preserved For The Future

A Tribute to South Wales Miners
Teyrnged i Lowyr De Cymru

A permanent reminder of the big wheel and the winder
And the men who risked their lives to hew the coal,
The hardships and the tragedies for miners and their families,
Explosions, falls and gas that took their toll
And wrecked and maimed the body, not the soul.
This monument – a miner's lamp to show we'll not forget
Their fortitude and suffering, their spirit, toil and sweat.
These giants of our yesterdays, whose struggles time cannot erase,
Their sacrifice and courage we extol.

Hawys Glyn James

On Saturday, 13 May 2000 poet and songwriter, Hawys Glyn James read out the above tribute she had composed for the unveiling of the Coalfield Memorial of a miner's lamp with an eternal flame dedicated to the thousands of miners who died in the coalmines of the South Wales Valleys. The congregation sang 'Cwm Rhondda' accompanied by the Lewis Merthyr Band.

Cwm Rhondda

Guide me, O! Thou great Jehovah,
Pilgrim through this barren land,
I am weak, but Thou art mighty,
Hold me with Thy powerful hand,
Bread of Heaven,
Feed me now and evermore

Open now the crystal fountain
Whence the healing streams do flow,
Let the fiery cloudy pillar
Lead me all my journey through:
Be Thou still my strength and shield.

Arglwydd arwain trwy'r anialwch,
Fi, bererin gwael ei wedd,
Nad oes ynof nerth na bywyd,
Fel yn gorwedd yn y bedd:
Hollalluog
Ydyw'r Un a'm cwyd i'r lan

Agor y ffynhonnau melys
Sydd yn tarddu o'r graig i maes
Hyd yr anial maith canlyned
Afon iachawdwriaeth gras:
Rho im hynny
Dim i mi ond dy fwynhau.

The colliery hooter closed the ceremony and dedications.

Black Gold, the Story of Coal
Journey back in time to experience the character and culture of the Rhondda, as seen through the eyes of three generations of one local mining family.

The Energy Zone
Join Trefor and Bertie in the Energy Zone for a fun, action-packed adventure – climb the temple of the Sun, mine your own energy in Castle Coal and discover how coal was made.

A Shift in Time.
Take the trip of your lifetime on the underground tour. Experience for yourself the hardship and joys of life underground in Lewis Merthyr Colliery.

I know that when you leave, you will go away having experienced a warm Rhondda welcome, a fond farewell, and a wish to come back again. See you soon!

When Coal was King Rhondda Reigned Supreme
The Rhondda Heritage Park in the South Wales Coalfield